TOWARD A SUSTAINABLE
WINE INDUSTRY
Green Enology in Practice

TOWARD A SUSTAINABLE WINE INDUSTRY

Green Enology in Practice

Edited by
Luann Preston-Wilsey

AAP | APPLE ACADEMIC PRESS

Apple Academic Press Inc. | Apple Academic Press Inc.
3333 Mistwell Crescent | 9 Spinnaker Way
Oakville, ON L6L 0A2 | Waretown, NJ 08758
Canada | USA

©2015 by Apple Academic Press, Inc.

First issued in paperback 2021

Exclusive worldwide distribution by CRC Press, a member of Taylor & Francis Group

No claim to original U.S. Government works

ISBN 13: 978-1-77463-546-9 (pbk)
ISBN 13: 978-1-77188-125-8 (hbk)

Library and Archives Canada Cataloguing in Publication

Toward a sustainable wine industry: green enology in practice/edited by Luann Preston-Wilsey.

Includes bibliographical references and index.
ISBN 978-1-77188-125-8 (bound)
1. Wine industry--Environmental aspects. 2. Wine and wine making--Environmental aspects. 3. Sustainable agriculture. I. Preston-Wilsey, Luann, editor

HD9370.5.T69 2015   338.4'76632   C2014-907762-9

Library of Congress Cataloging-in-Publication Data

Toward a sustainable wine industry: green enology in practice / editor: Luann Preston-Wilsey. -- 1st edition.

pages cm
Other title: Green enology in practice
Includes bibliographical references and index.
ISBN 978-1-77188-125-8 (alk. paper)
1. Wine and wine making--Environmental aspects. 2. Wine industry. 3. Green technology. I. Preston-Wilsey, Luann, editor. II. Title: Green enology in practice.

TP548.T68 2015   338.4'76632--dc23   2014045333

Apple Academic Press also publishes its books in a variety of electronic formats. Some content that appears in print may not be available in electronic format. For information about Apple Academic Press products, visit our website at **www.appleacademicpress.com** and the CRC Press website at **www.crc-press.com**

ABOUT THE EDITOR

LUANN PRESTON-WILSEY

Luann Preston-Wilsey of Cornell University possesses nearly two decades of professional winemaking experience. She is currently the technical supervisor at the Vinification and Brewing Laboratory at Cornell University's New York State Agricultural Experiment Station, at the Food Research Laboratory. Preston-Wilsey conducts trials with faculty researchers and industry leaders to evaluate grape breeding selections, viticulture treatments, and enological practices. She has carried out thousands of research fermentations, investigating such winemaking parameters as yeast and bacteria interactions, yeast nutrition, and tannin management.

CONTENTS

ACKNOWLEDGMENT AND HOW TO CITE

The editor and publisher thank each of the authors who contributed to this book. The chapters in this book were previously published in various places in various formats. To cite the work contained in this book and to view the individual permissions, please refer to the citation at the beginning of each chapter. Each chapter was read individually and carefully selected by the editor; the result is a book that provides a nuanced look at sustainable enology. The chapters included examine the following topics:

- Chapter 1 examines practical and concrete methods to assess the degree and quality of a winery's sustainable business practices. Importantly, it includes both direct and indirect emissions in its measurements ecological footprint.
- In Chapter 2, we switch from looking at environmental issues from the winery perspective to analyzing the footprint of a single bottle of red wine shipped from Australia to the UK. The authors' conclusions can be generalized across the board for green enology: after viticulture, transport lays the largest claim on the Earth's resources, making a good argument for drinking local wine. Packaging is also a significant factor, encouraging recycling and bulk packaging.
- In Chapter 3, the fermentation kinetics and metabolic compounds produced by multistarters during fermentation of organic musts are compared, providing a foundation for further research into organic yeasts, a vital factor in green enology.
- Chapter 4's wastewater research has a gate-to-grave perspective that includes viticulture factors, but some of the strategies can be applied to enology phases as well.
- The authors of Chapter 5 acknowledge that the closure impacts are only one small element of the overall system, but they also stress that their research is important because it is only by optimizing the thousands of products and product parts that we are using in our everyday lives that we can achieve sustainability.

- Although the characteristics of wastewater produced by wineries differ immensely from winery to winery, the authors of Chapter 6 offer concrete practices that wineries can implement to reduce the volume and the pollution load.
- Chapter 7 makes clear that any eco-labeling initiative needs to deliver an awareness of "premium" to consumers. Otherwise, consumers tend to feel that green enology is morally a "good" thing to support, but that it delivers wine of lesser quality.
- Chapter 8 studies young-adult consumers' willingness to pay for organic and sustainable wine.
- The authors of Chapter 9 contribute to the stakeholder literature that indicates stakeholders will be more likely to adopt green practices when it's to their benefit in some way by showing that future generations should be considered as a main stakeholder. Family-owned businesses are more likely to be favorable toward eco-certification because of the perceived long-term benefits.
- Although Chapter 10 does not focus specifically on wineries, it lists the factors that are important to create an integrated sustainable business, discussing issues that will be important for wineries to consider as they commit to being green industries.

LIST OF CONTRIBUTORS

Irene Aguzzi
Department of Food Science, University of Teramo, Teramo, Italy

David Amienyo
Sustainable Industrial Systems Group, School of Chemical Engineering and Analytical Science, The University of Manchester, Room C16, The Mill, Sackville Street, M13 9PL, UK

Adisa Azapagic
Sustainable Industrial Systems Group, School of Chemical Engineering and Analytical Science, The University of Manchester, Room C16, The Mill, Sackville Street, M13 9PL, UK

Cecil Camilleri
The Yalumba Wine Company, Angaston, SA 5353, Australia

Adolfo Carballo Penela
Fisheries Economics and Natural Resources Research Group-USC, Facultade de CC Económicas e Empresariais, Av. Burgo de las Nacións s/n, Santiago de Compostela, CP. 15782, A Coruña-Galicia, Spain

María do Carme García-Negro
Fisheries Economics and Natural Resources Research Group-USC, Facultade de CC Económicas e Empresariais, Av. Burgo de las Nacións s/n, Santiago de Compostela, CP. 15782, A Coruña-Galicia, Spain

Rachel J. C. Chen
Center for Sustainable Business and Development, 311 Conference Center Building, The University of Tennessee, Knoxville, TN 37996

T. E. Cloete
Department of Microbiology, Stellenbosch University, Private Bag X1, Matieland 7602, South Africa

A. Conradie
Institute for Wine Biotechnology, Stellenbosch University, Private Bag X1, Matieland 7602, South Africa

Magali Delmas
UCLA Institute of the Environment and Sustainability and Anderson School of Management, LaKretz Hall, Suite 300 Los Angeles, CA 90095

Juan Luís Doménech Quesada
Port of Gijón, c/Claudio Alvargonzález, 32; 33290 Gijón, Asturias, Spain

Elizabeth Duarte
Technical University of Lisbon, Instituto Superior de Agronomia

Giuseppe Fasoli
Department of Food Science, University of Teramo, Teramo, Italy

Olivier Gergaud
2KEDGE—Bordeaux Business School, Talence, France

Sebastien Humbert
Quantis, Parc Scientifique EPFL, Bâtiment D, 1015 Lausanne, Switzerland

Olivier Jolliet
Quantis, Parc Scientifique EPFL, Bâtiment D, 1015 Lausanne, Switzerland

Anna Kounina
Quantis, Parc Scientifique EPFL, Bâtiment D, 1015 Lausanne, Switzerland and Swiss Federal Institute of Technology Lausanne (EPFL), 1015 Lausanne, Switzerland

Neil Lessem
UCLA Institute of the Environment and Sustainability and Economics, LaKretz Hall, Suite 300, Los Angeles, CA 90095

Yves Loerincik
Quantis, Parc Scientifique EPFL, Bâtiment D, 1015 Lausanne, Switzerland

Rosa Maria Marianella
Direzione Generale della Prevenzione e Repressione Frodi Roma, Ministero delle Politiche Agricole Alimentari e Forestali, Roma, Italy

Jean-François Ménard
Quantis, 395 rue Laurier Ouest, Montréal, Québec, H2V 2K3, Canada

Margarida Oliveira
Instituto Superior Politécnico de Santarém, Escola Superior Agrária de Santarém and Technical University of Lisbon, Instituto Superior de Agronomia

Richard Pfister
Praxis Energia, rue Verte, 1261 Le Vaud, Switzerland

Amanda Pike
Quantis, 283 Franklin St. Floor 2, Boston, MA 02110, USA

Maria Schirone
Department of Food Science, University of Teramo, Teramo, Italy

Manuel Sergi
Department of Food Science, University of Teramo, Teramo, Italy

G.O. Sigge
Department of Food Science, Stellenbosch University, Private Bag X1, Matieland 7602

Giovanna Suzzi
Department of Food Science, University of Teramo, Teramo, Italy

Elisa Tatti
Quantis, Parc Scientifique EPFL, Bâtiment D, 1015 Lausanne, Switzerland

Rosanna Tofalo
Department of Food Science, University of Teramo, Teramo, Italy

Riccardo Vecchio
University of Naples Federico II, Department of Agricultural Sciences via Università, 100 80055 Portici, Italy

INTRODUCTION

INTRODUCTION: WHAT IS GREEN WINEMAKING?

The wine industry relies upon a connection to the ecosystem because it is first and foremost an agricultural enterprise. Wineries also use green vineyards and sparkling lakes as a large portion of their image and marketing appeal to customers. This multi-faceted natural relationship carries with it the responsibility to preserve and protect these resources so the wine is of the highest possible quality and tourists continue to be attracted to the setting. The so-called "locavore" movement has been a great blessing to many sectors of food production, and wineries across the globe have been among the greatest beneficiaries. Connected to this local movement is not only a re-evaluation of place, but also of production methods. When it comes to food and beverages, modern customers are increasingly interested in the "how" as much as the "where."

"SUSTAINABLE" VS. "ORGANIC" VS. "BIODYNAMIC"

In many winemaking regions, the government regulates the use of the term "organic," but "sustainable" and "biodynamic" have no legal definitions. When it comes to wine, consumers are often confused by these terms and unsure of what the terms mean. Even within the already listed categories, variations exist.

Wines can be made from grapes that have been certified organically grown, avoiding any synthetic pesticides or fertilizers. The European Union certifies these wines as "organic"; in the United States, however, the USDA will only allow the "organic wine" label if the wine is both made from organically grown grapes and is made without any added sulfites (although naturally occurring sulfites will still be present). (See Appendix for the range of organic definitions.)

Sustainable winemaking, on the other hand, includes practices that are not only ecologically sound but also both economically viable and socially responsible. Sustainable wineries will focus on energy and water conservation; they may use renewal energy sources. They will reduce air emissions, as well as wastewater that might cause soil and water pollution. They may also produce organic wine. Sustainability certifications are offered by a few third-party organizations, and regional industry associations are working on developing clearer standards than currently exist.

Biodynamic wine refers to how the grapes were grown before they were made into wine, as well as what goes on inside the winery. The vineyard is considered an ecosystem (and viewed from a holistic perspective that includes factors such as soil, other plants and wildlife, weather and climate, and seasonal cycles). The winemaker doesn't use traditional manipulations such as yeast additions or acidity adjustments. As with the term "organic," a wine might also be "made from biodynamic grapes," meaning that the winery used biodynamically grown grapes but then followed less strict rules for winemaking.

All three terms could be considered aspects of green enology.

Luann Preston-Wilsey

Corporate carbon footprint (CCFP) is one of the most widely used indicators to synthesise environmental impacts on a corporate scale. In Chapter 1, Carballo Penela and colleagues present a methodological proposal for CCFP calculation on the basis of the "method composed of financial accounts" abbreviated as MC3, considering the Spanish version "metodo compuesto de las cuentas contables". The main objective is to describe how this method and the main outputs obtained work. This latter task is fulfilled with a practical case study, where the authors estimate the carbon footprint of a wine-producing company for the year 2006. Results show the origin of impacts generated, providing this firm with disaggregated information on the contribution to its CCFP of each one of its activities and consumptions.

The UK consumes almost 5% of world's wine production, drinking 12.9 million hectolitres annually or 21 l per capita per year. Australian wines are most popular with the UK consumer, accounting for around 17% of total take-home purchases. Chapter 2, by Amienyo and colleagues, focuses on Australian red wine and presents the life cycle environmental impacts of its consumption in the UK. The results indicate that a 0.75 l bottle of wine requires, for example, 21 MJ of primary energy, 363 l of water and generates 1.25 kg of CO_2 eq. For the annual consumption of Australian red wine, this translates to around 3.5 PJ of energy, 600 million hectolitres of water and 210,000 t CO_2 eq. Viticulture and wine distribution are the main hot spots contributing over 70% to the environmental impacts considered. Shipping in bulk rather than bottled wine would reduce the global warming potential (GWP) by 13%, equivalent to 27,000 t CO_2 eq. annually. For every 10% increase in recycled glass content in bottles, the GWP would be reduced by 2% or 3600 t CO_2eq./yr; the savings in other environmental impacts are smaller (0.7–1.5%). A 10% decrease in bottle weight would reduce the impacts by 3–7%; for the GWP, the saving would be 4% or 7000 t CO_2 eq./yr. If only 10% of the wine was packaged in cartons instead of glass bottles, the GWP savings would be 5% or 10,600 t CO_2 eq./yr; the other impacts would also be reduced by 2–7%. These measures could together save at least 48,000 t CO_2 eq./yr, almost a quarter of the current emissions from the UK consumption of Australian red wine.

In the last years the use of a multistarter fermentation process has been proposed to improve the organoleptic characteristics of wines. In Chapter 3, Suzzi and colleagues investigated the fermentation performances and the interactions of mixed and sequential cultures of *Hanseniaspora uvarum*, *Candida zemplinina*, and a strain of *Saccharomyces cerevisiae* isolated from organic musts. To evaluate the oenological performances of the tested strains microvinifications in pasteurized red grape juice from Montepulciano d'Abruzzo cultivar were compared. The course of fermentation has been controlled through classical determinations (CO_2 evolution, ethanol, glycerol, pH, total titratable acidity, sugar content, free sulfur dioxide (SO_2), dry extract, sugars, organic acids, and volatile compounds). Moreover, the yeast population was determined by both culture-dependent and independent approaches. In particular, the pure culture of *H. uvarum* and *C. zemplinina* did not end the fermentation. On the contrary, when

S. cerevisiae was added, fermentations were faster confirming that yeast interactions influence the fermentation kinetics. Moreover, *C. zemplinina* showed a good interaction with *S. cerevisiae* by increasing the fermentation kinetic in high gravity Montepulciano must, with low ethyl acetate and acetic acid production. This study confirmed that non-*Saccharomyces* yeasts play a crucial role also in organic wines and their activity could be modulated through the selection of appropriate strains that correctly interact with *S. cerevisiae*.

The main goals of Chapter 4, by Oliveira and Duarte, are the comparison of different biological treatment systems, in particular fixed and suspended biomass, operating under aerobic conditions. Since the accurate design of the bioreactor is dependent on many operational parameters, aspects related to hydraulic retention time; oxygen mass transfer and contact time, energetic costs; sludge settling and production; response time during startup, flexibility and treated wastewater reuse, in crop irrigation, with the aim of closing the water cycle in the wine sector, will be addressed. A new treatment system will be presented as a case study, an air micro-bubble bioreactor (AMBB), that will highlight the advantages and constraints on its performance at bench-scale and full-scale, in order to fulfill the gaps associated with the implemented winery wastewater treatment systems. The data presented was collected during four years monitoring plan and used to develop a tool to support the selection of the best available technology. The present study will also contribute to the implementation of an integrated strategy for sustainable production in the wine sector, based on a modular and flexible technology that will facilitate compliance with environmental regulations and potential reuse for crop irrigation. This approach will contribute to the development of a bio-based economy in the wine sector that should be integrated in a Green Innovation Economy Cycle.

The objective of Kounina and colleagues in Chapter 5 was to discuss the implications of product loss rates in terms of the environmental performance of bottled wine. Wine loss refers to loss occurring when the consumer does not consume the wine contained in the bottle and disposes of it because of taste alteration, which is caused by inadequate product protection rendering the wine unpalatable to a knowledgeable consumer. The decision of whether or not to drink the wine in such cases is guided by subjective consumer taste perception and wine quality expectation (drinking the bottle or disposing of the wine down the drain and replacing it with a new bottle). This study aims to illustrate the importance of accurately defining system boundaries related to wine packaging systems. Methods: The environmental impacts resulting from wine loss rates as related to two types of wine bottle closures—natural cork stoppers and screw caps—have been estimated based on literature review data and compared to the impact of the respective closure system. The system studied relates to the functional unit "a 750 mL bottle of drinkable wine" and includes bottled wine, bottle and closure production, wine production, wine loss and wine poured down the drain. Results: The range of wine alteration rates due to corked wine is estimated to be 2-5% based on interviews with wine experts. Consumer behavior was assessed through a sensitivity study on replacement rates. When the increase in loss rate with the cork stopper is higher than 1.2% (corresponding to 3.5% corked wine multiplied by a consumer replacement rate of 35%), the influence of losses on the impact results is higher than that of the closure material itself. The different closures and associated wine losses represent less than 5% of the total life cycle impact of bottled wine.

The winemaking industry produces large volumes of wastewater that pose an environmental threat if not treated correctly. The increasing numbers of wineries and the demand for wine around the world are adding to the growing problem. The vinification process includes all steps of the winemaking process, from the receipt of grapes to the final packaged product in the bottle. To fully understand all the aspects of winery wastewater it is important to know the winemaking processes before considering possible treatments. Winemaking is seen as an art and all wineries are individual, hence treatment solutions should be different. Furthermore, wastewater also differs from one winery to another regarding its volume and composition and therefore is it vital for a detailed characterisation of the wastewater to fully understand the problem before managing it. However, prevention is better than cure. In Chapter 6, Conradie and colleagues describe a number of winemaking practices that can help lower the volume of the wastewater produced to decrease the work load of the treatment system and increase the efficiency of treatment.

Eco-labels are part of a new wave of environmental policy that emphasizes information disclosure as a tool to induce environmentally friendly behaviors by both firms and consumers. Eco-labels are often developed by third-parties separate to the industries that produce and sell the eco-product to create credibility. The goal of these agencies is to reduce the information asymmetry between producers and consumers over the environmental attributes of a good. However, by focusing on this information asymmetry, rather than how the label meets consumer needs, agencies may develop eco-labels that send an irrelevant, confusing or detrimental message to consumers. In Chapter 7, Delmas and Lessem use a discrete choice experiment to examine two similar eco-labels for wine, one associated with a quality reduction and the other not. The majority of respondents in our study were unaware of the difference between the labels. The authors found that respondents preferred eco-labeled wines over an otherwise identical counterpart, when the price was low and the wine was from a low quality region. However these preferences were reversed if the wine was expensive and from a high quality region. These results indicate that respondents obtain some warm glow value from eco-labeled wine, but also interpret it as a signal of low quality. This provides a clear lesson for policy makers that focusing purely on information asymmetries will not necessarily create eco-labels that align eco-products with the needs of consumers.

In Chapter 8, Vecchio explored young adult wine drinkers' willingness to pay (WTP) for three sustainable wines through Vickrey fifth-price full bidding auctions. In order to investigate factors affecting WTP the study compared the bid functions estimated with Tobit models and the premium functions estimated with ordinary least squares (OLS). The econometric results reveal that female and older respondents tend to bid higher for sustainable wines. Moreover, knowledge of specific claims increased price premiums. The findings have significant marketing and policy implications for the promotion of sustainable wines among young adults.

Business sustainability has been defined as meeting current needs while providing the ability of future generations to meet their own needs. However, few firms invest in practices geared at sustainability. In Chapter 9, Delmas and Olivier investigate how family ties to future generations via the intention of transgenerational succession can be associated with the adoption of sustainable practices. Using data from 281 wineries in the United States collected through a survey questionnaire, the authors show that ties to future generations, measured as the intention of the winery owner to pass down the winery to their children, are associated with the adoption of sustainable certification.

We conclude this compendium with a look at the factors that need to be considered when building any sustainable business. Author Rachel Chen stresses the importance of businesses monitoring and managing their environmental impacts, while they aim to integrate, with consistent quality control, effective reduce-reuse-recycle programs and risk preventions. By building an integrated sustainable business and development system to meet certain environmental standards, many companies are eligible to be "green" certified. Companies may consider recognizing global visions on sustainability while implementing local best practices. An integrated sustainable business and development system includes talent management, sustainable supply chain, practicing strategies of leveraging resources effectively, implementing social responsibilities, initiating innovative programs of recycling, reducing, and reusing, advancing leaders' perceptions towards sustainability, reducing innovation barriers, and engaging sustainable practices strategically.

PART I

WINERY ASSESSMENTS

CHAPTER 1

A METHODOLOGICAL PROPOSAL FOR CORPORATE CARBON FOOTPRINT AND ITS APPLICATION TO A WINE-PRODUCING COMPANY IN GALICIA, SPAIN

ADOLFO CARBALLO PENELA,
MARHA DO CARME GARCHA-NEGRO,
AND JUAN LUHS DOMÉNECH QUESADA

1.1 INTRODUCTION: CORPORATE SUSTAINABILITY AND ECOLOGICAL FOOTPRINT ANALYSIS

Over the last few decades, organisations have gained better awareness on issues that have traditionally played a secondary role or were simply not regarded as business strategies. This is particularly true in the case of environmental sustainability-related questions.

A Methodological Proposal for Corporate Carbon Footprint and Its Application to a Wine-Producing Company in Galicia, Spain. © Carballo Penela A, do Carme García-Negro M, and Doménech Quesada JL. Sustainability, *1 (2009), doi:10.3390/su1020302. Licensed under Creative Commons Attribution 3.0 Unported License, http://creativecommons.org/licenses/by/3.0. Used with permission from the authors.*

From a business perspective, more value is now placed on these matters, particularly in light of: i) the development of legislation related to emissions and waste discharge control levels; ii) informational tasks regarding environment-related issues; iii) the demand for higher transparency and commitment by related agents; iv) an awareness that the relation with several collectives (shareholders, clients, workers, community, etc.) and the environment is integrated with the firm's value; or v) the search for new tools to manage profits and risks derived from intensifying globalisation processes and the disappearance of national borders.

Companies have realised that sustainability constitutes a means of differentiation, which is crucial for increasing productivity and competitiveness. The adoption of proactive sustainability management has direct and positive repercussions on business's competitiveness [1].

The concept of Corporate Social Responsibility (CSR) and the appearance of guidelines and agreements to standardise the design and implementation of reports on an organisation's environmental, social, and economic performance are good examples of the concern for incorporating an environmental perspective in business management. Several proposals in this sense, such as the "Global Report Initiative" (GRI) have gained recognition worldwide [2,3]. However, even if these initiatives constitute important achievements, its use is left up to organisations, as well as the way in which the chosen ones are applied. This is due to the lack of strict recommendations in this sense.

This reflects a highly important issue: the measure of an organisation's environmental performance. Therefore, different standards usually suggest a list of specific indicators, which can sometimes offer contradictory results. This poses difficulties in diagnosing and reporting results.

Authors such as Holland [4] have pointed out the non-existence of instruments that can synthesise organisations' environmental situation using a holistic approach that can be used for decision-making and communication with shareholders, stakeholders, and society in general.

In this context, several papers [5-11] deal with the possibility of applying EF analysis to companies and their products, contributing ideas or developing different methodologies on measuring corporate ecological footprint (CEF).

The EF is an indicator that assesses the demand for biocapacity by inhabitants of a geographical area to maintain their consumption of resources and generation of wastes using existing technology [12]. Comparing the EF of a given territory with its available biologically productive space (BPS) enables the determination of the extent to which its carrying capacity is exceeded. This indicator dates back to the early 1990s [13,14], and has undergone different changes that solved certain initial deficiencies [15-19].

We are dealing with a versatile tool that is capable of being applied to contexts other than populations, such as organisations, products, and different kinds of activities [14]. Chamber and Lewis and Holland [4,5] report some of the contributions of this indicator towards the achievement of sustainability by firms and organisations. First, this single index synthesizes different environmental impacts and allows an evaluation of the success or the failure of the measures adopted. Second, its calculation methodology is relatively simple. Third, this indicator is expressed in easily understandable units, making decision-making and internal and external reporting manageable. Fourth, the information needed is based on data available in company records.

Some doubts could be cast on the sense of using an indicator that is expressed in hectares of productive space, a unit that is appropriate for countries or regions, but maybe less related to corporations. However, the different types of land distinguished by EF also provide information relevant for companies, expressing in a common—and therefore, agreeable—unit the influence of issues such as direct and indirect energy consumption, wastes generation, etc. [4,20].

On the other hand, use of the so-called carbon footprint (CFP) is becoming increasingly widespread. CFP is a footprint that measures CO_2 or other greenhouse gas emissions. As happens with EF, this indicator can be applied to companies and organisations, with the concepts of corporate carbon footprint (CCFP) being a very attractive indicator at this level, especially when we consider the demands in the framework of the Kyoto Protocol.

Different approaches have been used to estimate CCFP [5,9,21-24]. However, there is still no consensus regarding certain matters that determine its content (e.g., the inclusion of CO_2 or other gas emissions), scope (only direct emissions or also indirect emissions, being embodied in the purchase of goods and services that need energy in their production), and methodology (classical life-cycle analysis, input–output techniques, etc.). In this paper, we offer a methodological alternative for CCFP, the "method composed of financial accounts" (MC3), which is described in Section 2. Next, we apply MC3 to a wine-producing firm, which enables us to verify the practical utility of the selected method (Section 3). Finally, we discuss the main conclusions obtained (Section 4).

1.2 A METHODOLOGICAL PROPOSAL FOR CCFP CALCULATION: THE METHOD COMPOSED OF FINANCIAL ACCOUNTS (MC3)

1.2.1 THE MC3 FOUNDATIONS

The calculation method used, MC3, has been developed by Doménech [9,20]. Doménech starts from the need to make a method that enables the estimation of companies' and organisations' CEF, offering the possibility of expressing this footprint both in land units and in t CO_2, so that CCFP can be calculated.

The origin of MC3 can be found in the concept of household footprint [25]. In this way, based on the matrix of consumptions versus land present in the spreadsheet for the calculation of households' footprint by Wackernagel [25], Doménech [9] prepares a similar matrix, which contains the consumptions of the main categories of products needed by a company, and also includes sections for the wastes generated and the use of land. These consumptions/wastes will be transformed into land units and t CO_2.

MC3 was first applied to the Gijón Port Authority [9]. Later on, it was tested and improved by the Working Group on Corporate Ecological Footprint Enhancement, coordinated by Doménech himself, in which five Spanish universities took part. For a year and a half, this method has been applied to firms belonging to different economic sectors [26,31]. This application period proved this method is robust and useful in providing information relevant to improve companies' and organisations' environmental performance in any economic sector.

The CCFP obtained with the current MC3 version includes direct and indirect CO_2 emissions, the latter being considered as those generated in the production/provision of goods and services obtained. A second version of this method is currently under development, which will incorporate emissions from the rest of the greenhouse gases included in the Kyoto Protocol, by using the GWP coefficients with a time horizon of 100 years prepared in [32]. These coefficients relate each gas-warming potential with CO_2-warming potential, which makes the needed conversion possible. For example, a factor 23 means that the contribution by gas unit is 23 times higher than that from CO_2.

Hence, this indicator will be expressed in CO_2 equivalent tons. Furthermore, emissions derived from the use of land capable of sequestering CO_2 in the same way as forest (pastures, croplands, etc.) will also be incorporated in the CCFP. Nonetheless, this paper describes and applies the initial version of the method. The information necessary to estimate CCFP using MC3 is mainly obtained from accounting documents such as the balance sheet and the income statement. Hence, the denomination "method composed of financial accounts" (MC3).

However, further information from other company departments with specific data about certain sections (waste generation, use of land by the organisation's facilities, among others) may also be necessary. The footprint is calculated in a spreadsheet, which is used not only to estimate CCFP, but also to determine CEF. This spreadsheet works as the land-use matrix used for the calculation of countries' EF. Besides showing results, this spreadsheet also makes the estimation of both indicators possible.

1.2.1.1 CORPORATE LAND-USE MATRIX

The rows of this matrix show the footprint of each category of product/service consumed. The columns present, among other elements, different land-use categories, into which the footprint is divided (see Table 1).

Columns are divided into six groups. The first one (column 1) corresponds to the description of the different categories of consumable products. These are classified into four major categories: energy consumption, which is subdivided into six subgroups (electricity, fuels, materials, construction materials, services, and waste products), use of land, agricultural resources, fishing resources, and forest resources. One can include as many products as desired within each group.

The second group (columns 2-6) shows each consumption of product, expressed in specific units. The units in the first column of the group are related to product's characteristics (electricity consumption is expressed in kwh, water in m^3...). The second column indicates the value of consumptions in monetary units, while the third shows consumptions in tons. The fifth column reveals energy corresponding to each consumption expressed in gigajoules (GJ), which is obtained by multiplying tons of product by the quantity of energy used by ton in its production (GJ/t). The result is called energy intensity, indicated in the fourth column.

The third group of columns (columns 7 and 8) show each good's productivity, in a way that one column indicates natural productivity, expressed in tons per hectare, while another presents energy productivity, expressed in GJ per hectare.

The fourth group is composed of six columns (9-14) showing the distribution of the footprint among different categories of land. These are the same as that used for the countries' EF (fossil energy, cropland, pastures, forests, built-up land, and sea).

There is another group (columns 15 and 16), which collects the total ecological footprint (i.e. occupied land) and counterfootprint, that is, available land. The counterfootprint concept will be described in subsequent sections.

1.2.1.2 CCFP CALCULATION

As discussed earlier, the methodology developed by Doménech is thought to calculate both CCFP and CEF. In this section, we describe the calculation processes of both indicators, stressing the peculiarities of CCFP calculation.

The design of this calculation method starts from the EF philosophy. Nevertheless, there are important differences between an organisation and a population, which reflects in the development of a specific calculation method to study the firms' footprint.

Many of the goods consumed by a company do not come directly from any BPS. Companies purchase machinery and computers, consume electricity, contract services etc., and the ecological footprint of all these consumptions cannot be calculated by dividing the consumption by the productivity of the productive space, from which these goods come, because they are not biotic and therefore do not have any direct origin.

As a consequence, a problem arises, because many goods and services cannot be included in CEF in the usual way. In this case, the method philosophy is similar to that adopted in the study for territories' footprint, since in addition to direct energy consumption, we show the impact of the energy used in the production of goods and services consumed by the organisation studied. These consumptions are precisely those originating from most CCFP.

In the case of territories, the total energy consumption by the inhabitants of a country or region under study is taken into account, making an additional adjustment depending on its goods imports or exports. Since this is not possible with firms, Doménech uses energy intensity factors, which indicate the energy consumed in the production of each product category, expressed in gigajoules per ton. These energy intensity factors would be of the same type as those used in the calculation of countries' footprints to determine the quantity of energy incorporated to commercial flows.

Therefore, the fourth and fifth columns of the second group in the spreadsheet make sense, since we obtain the total energy incorporated in the production of each product multiplied by consumption, which is expressed in tons, by energy intensity (GJ/t). In the case of depreciable goods, CEF collects its depreciation quota each year, avoiding high fluctuations in the periods when fixed assets are acquired.

Firms' information on some consumption is hardly expressed in tons, while it is normally available in money expenditure or in some cases (fuels, electricity) other physical units, such as litres, kwh, etc. In the first case, the conversion into tons can be made by considering the specific product average prices in the period under study (for example, euros/kg). Another option is to use foreign trade statistics, which offer information about imports and exports of the different tariff chapters expressed in monetary units and tons, thus enabling one to obtain a monetary unit/ton factor. In the second case, the transformation is made considering the item specific weight. In the case of electricity, we consider the quantity of fuel used to obtain one kwh.

Energy intensity factors comprise the amount of energy used in the production of every product included in the corporate land-use matrix (e.g. fertilisers, industrial machinery, etc.), considering an average life cycle. They are obtained from Wackernagel [25,33] and other researchers' studies [34,35].

Regarding biotic or natural resources, whose consumption can be transformed into land in the normal way [14], CEF also includes the energy incorporated in their production, which is calculated along with the rest of the goods by applying an energy intensity factor to the consumption of agricultural, fishing, and forest resources.

Energy footprint is also estimated for services contracted by the organisation under study, as well as for the wastes it generates, both aspects being important for the organisations' footprint. In relation to the first, it is assumed that part of the service cost corresponds to energy consumption, and we estimate the weight of this part for each kind of service. After applying this percentage to the service price, we obtain "euros corresponding to the energy consumption" [20]. This value is transformed into tons, considering fuel prices. Afterwards, the corresponding energy intensity is applied, in the same way as when estimating the energy footprint of any other non-biotic resource.

The methodology for waste discharges and emissions is still under development. In the case of wastes, the footprint is based on the calculation of the energy consumed during wastes management. Here, the quantity of energy recovered in recycling processes may be discounted. When doing so, we do not register the possible damaging effects caused by wastes but the energy consumption they generate.

Thus, we obtain the total energy consumed by a given organisation by considering its direct energy consumption of electricity and fuels as well as the part that is indirectly consumed, which is already incorporated in the goods and services used by the firm and in the wastes it generates.

Once this done, the CEF still compares consumption with the quantity of energy that can be assimilated by a hectare of forest according to

CO_2 emissions; in other words, the energy productivity of each fuel is expressed in GJ/ha. In other words, we estimate how many gigajoules of each fuel were needed to emit the CO_2 volume that can be absorbed by a hectare annually, applying an absorption rate by hectare and year of 5.21 CO_2 t/ha/year [32]. For instance, an average world forest hectare can absorb e CO_2 emissions from a consumption of 71 GJ per year of liquid fuels or 55 GJ of coal [33].

CCFP calculation does not need to resort to these types of factors in this case and total energy consumption is easily transformed into t CO_2 by considering the emission factors that indicate the quantity of CO_2 emitted by GJ consumed in each type of fuel.

After this task, we have already obtained most CCFP. To complete it, the calculation method shows the emissions caused by the firm's consumption of forest resources (wood, rubber, paper, etc.). We must remember that an organisation may consume "biotic" resources, such as food and wood, which are directly associated with a type of BPS (croplands, pastures, forests, and sea). In this case, CEF not only includes the energy incorporated in obtaining these goods, estimated as we have specified, but also considers the productive space that is needed to make these consumptions. This

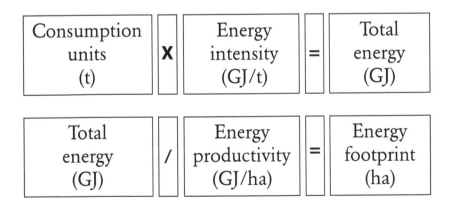

FIGURE 1: The energy footprint.

land is calculated in the normal way found in EF studies, by dividing the consumption of each product, expressed in this case in tons, by the natural productivity of the BPS assigned to each product. Natural productivity collects the amount of every natural resource that humans are able to extract per hectare. Natural productivity data are often obtained from FAO statistics (http://faostat.fao.org/site/339/ default.aspx). For instance, if we consume 10 tons of wood and forest productivity is 1.19 t/ha, the wood footprint assigned to forest land would stand at 8.40 ha, to which the corresponding emission factor should be applied.

In addition, consumption from forests contributes to reducing the forest's capacity to absorb CO_2. Hence, it is considered that non-absorbed emissions derived from forest products consumed by a firm must be included in its CCFP. In this manner, once we have determined the hectares of forest needed by the organisation under study, these hectares are multiplied by the 5.21 CO_2 t/ha absorption rate, to estimate how much CO_2 is no longer absorbed.

Finally, CEF applied to organisations shows the use of productive space, both on land and at sea. Thus, the types of land are differentiated (build-up, croplands, pastures), organisations' counterfootprint being also estimated.

The counterfootprint concept can be partially assimilated to the BPS of a country or region. In the traditional ecological footprint analysis, there is a comparison between the land needed to satisfy the needs of a given population, the EF itself, and the productive space available to satisfy these needs. From this comparison, we obtain either a deficit or an ecological reserve, depending on which of the two spaces is larger.

However, the BPS concept makes sense when dealing with territories, but not as much with organisations. All countries use, to a certain extent, part of their land to produce biotic resources. This is why the comparison between available and consumed land is always possible. EF assesses the availability of BPS, and consequently, the fact that a territory population satisfies its needs with products that originate in the territory itself. From the aspect of sustainability condition, a country without productive space

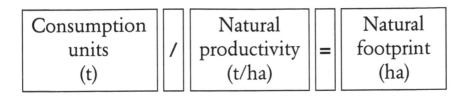

FIGURE 2: The natural footprint.

can hardly be sustainable, because its inhabitants always need to consume, even just to satisfy their vital needs.

In the case of companies, this assumption is difficult to maintain, since many of them do not need land to produce biotic resources. A car garage or a financial entity will develop their activities without any direct link with this type of resources. This is why the concept of counterfootprint appears. It starts from the positive regard for companies' availability of natural capital, despite the desirable reduction of their footprint by being more efficient and by curbing consumption.

Therefore, investments in this kind of productive space reduce their footprint. In this way, this indicator could encourage the private sector involvement in the preservation of natural spaces (Doménech, 2007), this being considered positive in terms of sustainability. Land devoted to crop-lands, pastures, forests, gardens, or for instance marine reserves owned by a firm will contribute to partially counteract CEF, since all these are considered counterfootprint. To reduce a hectare of footprint, it is necessary to acquire the same quantity of one of these spaces.

When investments are made in wooded land, CO_2 emissions and consequently CCFP will also decrease, considering the absorption rate of 5.21 CO_2 t/ha/year. Net CEF is obtained by subtracting counterfootprint from CEF. In the same way, net CCFP is the result of subtracting the CO_2 absorbed by investments in counterfootprint from CCFP.

1.2.2 THE MC3 AND OTHER METHODOLOGICAL APPROACHES FOR THE ESTIMATION OF CEF AND CCFP

Nowadays, approaches such as BL3 or the "component-based approach" (CBA) estimate CCFP and CFP. Even though the results obtained with these methods and MC3 are similar, there are relevant differences in terms of the calculation method and some assumptions involved in the estimation of the indicator: the CCFP of a given company will be quite different depending on the approach chosen. Some of these differences are linked with issues shown in Table 2.

TABLE 2: The MC3 and other s methodological approaches.

Concept	BL3	CBA	MC3
Calculation method	Input-output analysis/LCA	Component-based approach/LCA	MC3 is based on Compound-Method
Activities included in CCFP	All the activities	Relevant activities	All the activities
Transformation of financial information into mass unit data	No needed. The method uses monetary input-output coefficients	Needed. No explicit method	Needed. Explicit method
Equivalence and yield factors	Yes	No	Yes
Is the required software accessible?	No	No	Yes

The input-output analysis is a useful tool to estimate CCFP considering emissions along the supply chain. This method avoids double counting problems and truncation errors, providing comparable footprints [10]. However, it also has some limitations related to sector aggregation, the consumption of imported goods, and errors from the use of country monetary tables, instead of physical unit tables to elaborate input-output coefficients, used to estimate CCFP [37].

CBA estimates the footprint of some relevant components of an organisation resource consumption and waste generation, assessing the CCFP of every component using life-cycle data. Issues such as the completeness of the component list and the reliability of the life-cycle assessment for each component are relevant in terms of the accuracy of the analysis. This approach presents some limitations, given LCAs' boundary problems; the lack of accurate and complete information about products' life-cycles, problems of double-counting in the case of complex chains of production, and the large amount of detailed knowledge necessary for each analysed process [18].

The use of MC3 involves several advantages since:

1. It is a complete method, which collects the footprint from the consumption of all goods and services and wastes generated by a company.
2. It is based on Wackernagel and Rees' "compound-method", a solid well-known method for researchers in ecological footprint.
3. It is a technically feasible method. Its calculation does not require extensive expert staff inputs: everybody working with spreadsheets is able to calculate CCFP.
4. It is a transparent method. The spreadsheet and all the data needed for the estimation of CCFP, including energy intensity factors and productivities, are available for researchers at http://www.huellae-cologica.com.
5. It is a flexible method. The spreadsheet offers the researchers the possibility of adding/changing the factors employed for the estimation, according to the specific needs of their company.

1.3 CCFP ESTIMATION FOR A WINE PRODUCER IN GALICIA, SPAIN

The company considered for this study, which we will refer to as Gamma, is devoted to the production and marketing of Albariño wine. This is a family-run firm with just two workers, although Gamma hires temporal staff for the grape-harvesting season and related activities, besides relying on other relatives' collaboration. This company's vineyards occupy 2.5 ha, having produced 24,686 kg of grapes in 2006. Moreover, Gamma also buys 19,682 kg of grapes from other producers, which results in a total wine production of 30,000 L.

The information needed to estimate Gamma's CCFP was obtained in two phases. First, the person responsible for supplying data was given information on EF analysis, stressing the utility of this indicator from a corporate perspective.

Subsequently, a tentative survey guide was prepared. The instrument tries to include the main consumption categories needed to estimate its footprint, although new categories were later incorporated, showing specific consumptions that were initially not taken into account. Available information refers to the 2006 accounting period.

1.3.1 RESULTS AND DISCUSSION

The following tables and figures show the main results related to Gamma's CCFP. Table 3 shows the company's total CO_2 emissions (152.7 t CO_2). Although vineyards absorb CO_2, there are no differences between Gamma's gross and net emissions, because the present MC3 version only considers the counterfootprint from wooded land. MC3 2.0 considers the productive space occupied by vineyards as the counterfootprint.

TABLE 3: CFP and ratios related to Gamma (2006).

Concept	Ud.	Gamma
Gross CO_2 emissions	t	152.7
Net CO_2 emissions	t	152.7
Sold goods	t	27.9
Net CO_2 t/goods tons	t	5.47

These emissions are related to the quantity of goods sold, 5.47 CO_2 t having been determined as the emission per ton of wine produced by Gamma.

Considering the production of the company, 30,000 litres, the CCFP of a 75 cL bottle of wine is 3,817 g CO_2, nearly five times the footprint shown in previous similar studies [38,39] (i.e. 789 and 835 g CO_2 per 75 cL bottle). We want to remark that MC3 includes CO_2 emissions from the production of every good and service consumed by a company, directly or indirectly linked with the production of wine. The footprint of capital goods is estimated allocating the total amount of CO_2 generated in their production, considering the duration of their expectable life. Our study also considers emissions from the building process of the company's warehouse, as well as other goods not included in previously mentioned studies. High energy intensities for chemicals and glass by-products also contribute to increase Gamma's CCFP.

The potential usefulness of these results are related to a) the possibility of comparing them with CCFP from other companies within the same sector; b) the analysis of Gamma's CCFP evolution along the time; c) the use of this information to estimate the footprint of the wine produced and marketed by Gamma from a perspective that not only considers the CCFP of the firm itself, but also that of its supplier chain. In this way, this indicator could be used in the preparation of an ecolabel, offering information that consumers might consider when deciding on substitute products.

MC3 allows advancing in the CCFP origin. A first classification distinguishes among emissions from consumption (either direct or indirect), from energy (133.7 t CO_2), and those caused by forest resources consumption (18.95 t CO_2) (Table 4).

TABLE 4: Gamma's CCFP: types of footprint.

Type of Footprint	t CO_2	%
Fossil energy	133.7	87.6
Forest	18.95	12.4
Total	152.7	100

Likewise, we can determine what consumptions hide from each one of these categories. In terms of energy, Figure 3 shows the distribution of CO_2 tons generated by the Gamma's direct and indirect energy consumption. We have observed that in 67% of these emissions, 88.6 t CO_2, comes from the energy incorporated in the production of non-depreciable materials consumed. Second, the consumption of agricultural resources and specifically of grapes—this firm's raw material—generates 14.4 CO_2 tons or 12% of emissions.

In this case, we are dealing with a product with low energy intensity, 10 GJ/t, but from whose Gamma acquires an important volume (19.7 t). The third CO_2 generator from fossil energy consumption is depreciable material, which generates 11.3 t CO_2, 8% of the total. It is remarkable that depreciable and non-depreciable materials are separated. In the case of the former, consumption is associated with the depreciation quota of the specific year. As for the latter, the real consumption of goods is shown.

Thus, we notice that over 85% of the CO_2 associated with energy consumption as well as almost 75% of the total CCFP are generated by just three consumption categories. Other consumptions such as fuels (7.2 t CO_2) or electricity (5.9 t CO_2), with a priori more visible emissions, are less important in this company.

Given the importance of the emissions from materials, we consider it appropriate to advance further, indicating which are those with the highest contributions to Gamma's CCFP. In terms of agricultural resources, total emissions are associated with a single product (grapes); hence, more advances are not necessary.

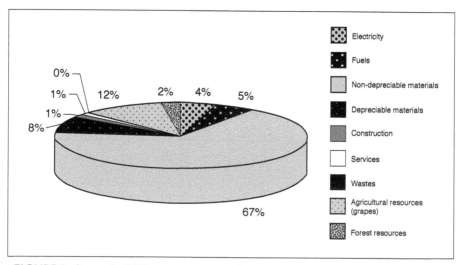

FIGURE 3: Gamma's CCFP: energy consumption.

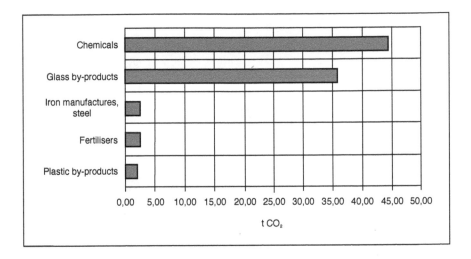

FIGURE 4: CCFP from non-depreciable materials (t CO_2).

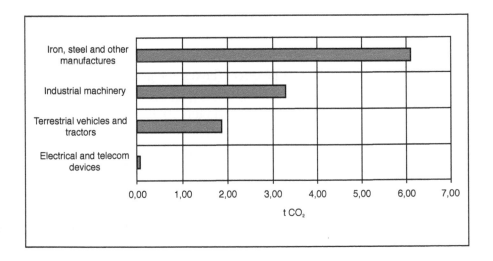

FIGURE 5: CCFP from depreciable materials (t CO_2).

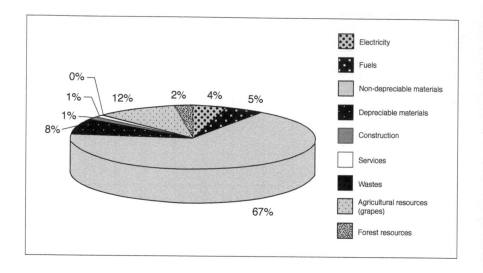

FIGURE 6: Gamma's CCFP: Consumption of forestry resources (t CO2).

Figure 4 shows the five non-depreciable materials with the highest CCFP. Chemicals (44.57 t CO_2) stand out, which include sulfates used by Gamma, and glass byproducts (35.70 t CO_2), result of consumption of bottles to bottle the firm's production. Emissions from other materials, such as iron and steel manufactures (mainly non-depreciable machinery), fertilisers, and plastic-derived products (boxes for grape-harvesting and synthetic furniture) rank next, although none of them reaches 3 t CO_2.

With regard to depreciable materials (Figure 5), their CCFP is considerably lower. In this respect, the 6.10 CO_2 t emission generated by the consumption of iron and steel manufactures is outstanding, which are related to the freezing tanks where wine is stored. Emissions generated from the production of specialised wine-making machinery (mainly bottling line, corking machines, and conveyor belts) reach 3.31 CO_2 tons, while emissions associated with other consumption concepts such as automobile vehicles and electrical devices are notably lower.

Emissions associated with forestry resources consumption (Figure 6) represent 12.4% of CCFP, reaching 18.95 t CO_2. In this case, 63% of emissions (11.77 t CO_2) comes from the consumption of cork products and their manufactures, in Gamma this being related to the purchase of cork for the 40,667 bottles needed for its 2006 production. Packing these bottles implies an important consumption of boxes and cardboard cases, making emissions associated with paper and cardboard consumption reach 5.9 t CO_2. Wood consumption, in Gamma's furniture and in the building of the firm's premises, does not reach 1 t CO_2.

1.4 CONCLUSIONS

The search for sustainability by organisations and companies is driven by the need to attain global sustainable development as well as a management tool to enhance a firm's value. Tools for assessing corporate environmental performance is a requisite for firms committed to the environment and to a view of business management beyond the traditional models.

The method composed of financial accounts (MC3) is presented as an alternative when the assessment of the firms' or organisations' sustainability based on their CCFP is included. This is compatible with well-known information standards, such as the "Sustainability Guidelines" of the *Global Reporting Initiative* (GRI). Moreover, this can be used to estimate a product's footprint, by adding the footprint corresponding to all firms, through which a given product circulates along its life cycle.

In this paper, we have applied MC3 to a company that we called Gamma, focusing in the calculation of its CCFP. We have verified that the information obtained is relevant for the design of measures targeted at the reduction of its CO_2 emissions. Results obtained indicate the activities generating the highest CCFP, which helps in determining the contribution of each consumption. Since both direct and indirect emissions are included, some activities appear whose impacts are not always easily noticed, a circumstance that increases the intervention possibilities.

Gamma should try to manage its materials in a more efficient way, reducing their consumption. This task may prove difficult on the short term for depreciable goods, since they are the firm's production means, and acquired to last for years. As for non-depreciable goods, the intervention margin is wider. In some cases (glass bottles or corks), it is difficult to reduce the number of units consumed without substantially affecting the firm's economic profitability. Despite this, a possible solution would be to try to purchase products with a lower weight per unit. In products such as fertilisers, for instance, the optimisation of consumption will contribute toward reducing Gamma's CCFP.

REFERENCES

1. An introduction to corporate environmental management: striving for sustainability, 1st ed.; Schaltegger, S., Burritt, R., Petersen, H., Eds.; Greenleaf Publishing: Sheffield, UK, 2003; pp. 1-384.
2. GRI (Global Reporting Initiative). Sustainability Reporting Guidelines; Available online: http://www.aeca.es/comisiones/rsc/documentos_fundamentales_rsc/gri/guidelines/gri_guidelines_2002.pdf (accessed October 23, 2006).
3. GRI (Global Reporting Initiative). Sustainability Reporting Guidelines. Available online: http://www.globalreporting.org/NR/rdonlyres/ED9E9B36-AB54-4DE1-BFF2-5F735235CA44/0/G3_GuidelinesENU.pdf (accessed November 23, 2008).
4. Holland, L. Can the Principle of the Ecological Footprint be Applied to Measure the Environmental Sustainability of Business? Corp. Soc. Responsibility Environ. Manage. 2003, 10, 224-232.
5. Chambers, N.; Lewis, K. Ecological Footprint Análisis: Towards a Sustainability Indicador for Business. ACCA Research Report No. 65, Oxford, UK, 2001.
6. Lenzen, M.; Lundie, S.; Bransgrove, G.; Charet, L.; Sack, F. Assessing the Ecological Footprint of a Large Metropolitan Water Supplier: Lessons for Water Management and Planning towards Sustainability. J. Environ. Plan. Manage. 2003, 46, 113-141.
7. Lenzen, M.; Foran, B.; Dey, C. Sustainability Accounting for Business - A new International Software Based on Input-Output Tables. Paper presented at the Intermediate Input-Output Meeting Conference, Sendai, Japan, 26-28, July, 2006.
8. Wiedmann, T.; Lenzen, M. Sharing Responsibility along Supply Chains - A New Life-Cycle Approach and Software Tool for Triple-Bottom-Line Accounting. Working paper presented at the Corporate Responsibility Research Conference, Trinity College Dublin, Ireland, 4-5, September, 2006.
9. Doménech, J.L. La huella ecológica empresarial: el caso del puerto de Gijón. Working paper preseted at the VIIth Nacional conference of environment, Madrid, Spain, 22-26, November, 2004.
10. Wiedmann, T.; Barret, J.; Lenzen, M. Companies on the Scale: Comparing and Benchmarking the Footprints of Businesses. Working paper presented at the International Ecological Footprint Conference, Cardiff University, Cardiff, Wales, 8-10, May, 2007.
11. Murray, J.; Dey, C. Assessing the Impacts of a Loaf of Bread; ISA Research Report 04-07: Sydney, Australia, 2007; pp. 1-43.

12. Wackernagel, M.; Monfreda, Ch.; Moran, D.; Wermer, P.; Goldfinger, S.; Deumling, D. Ecological footprint time series of Austria, the Philippines, and South Korea for 1961-1999: comparing the conventional approach to an actual land area' approach. Land Use Policy 2005, 21, 261-269.
13. Rees, W.E. Ecological Footprints and Appropriated Carrying Capacity: What Urban Economists Leaves Out. Environ. Urban. 1992, 4, 121-130.
14. Wackernagel, M.; Rees, W.E. Our Ecological Footprint: Reducing Human Impact on the Earth, 2nd ed.; New Society Publishers: Philadelphia, USA, 1996; pp. 1-160.
15. Wackernagel, M. The Ecological Footprint of Santiago de Chile. Local Environ. 1998, 3, 7-25.
16. Wackernagel, M.; Silverstein, J. Big Things First: Focusing on the Scale Imperative with the Ecological Footprint. Ecol. Econ. 2000, 32, 391-394.
17. Wackernagel, M.; Schulz, N.; Deumling, D.; Callejas Linares, A.; Jenkins, M.; Kapos, V.; Monfreda, C.; Loh, J.; Myers, N.; Norgaard, R.; Randers, J. Tracking the Ecological Overshoot of the Human Economy. Proc. Nat. Acad. Sci. 2002, 99, 9266-9271.
18. Monfreda, Ch.; Wackernagel, M.; Deumling, D. Establishing National Natural Capital Accounts Based on Detailed Ecological Footprint and Biological Capacity Assessment. Land Use Policy 2004, 21, 231-246.
19. Kitzes, J.; Peller, A.; Goldfinger, S.; Wackernagel. M. Currents Methods for Calculating National Ecological Footprint Accounts. Sci. Environ. & Sustain. Soc. 2007, 41, 1-9.
20. Doménech, J.L. Huella ecológica y desarrollo sostenible, 1st ed.; AENOR Ediciones: Madrid, Spain, 2007; pp. 1-398.
21. GFN (Global Footprint Network). Ecological footprint and biocapacity. Technical notes: 2006 ed., Oakland, CA, USA, 2006.
22. Wiedmann, T.; Lenzen, M. Unravelling the impacts of supply chains. A new Triple-Bottom-Line Accounting Approach; ISA UK Research Report 07-02: Durham, UK, 2007; pp. 1-26.
23. Wiedmann, T.; Minx, J. A definition of carbon footprint; ISA UK Research Report 07-01: Durham, UK, 2007; pp. 1-11.
24. BSI (Bristish Standards Institute). Specification for the assessment of the life cycle greenhouse emissions of goods and services. Available online: http://www.bsigroup.com/en/Standards-and-Publications/Industry-Sectors/Energy/PAS-2050/ (accessed November 10, 2008).
25. Wackernagel, M.; Dholakia, R.; Deumling, D.; Richardson, D. Assess your Household's Ecological Footprint 2.0. Available online: http://greatchange.org/ng-footprintef_household_evaluation.xls (accessed November 1, 2005).
26. Álvarez Díaz, P.D.; Doménech Quesada, J.L.; Perales Vargas-Machuca, J.A. Huella ecológica energética corporativa: Un indicador de la sostenibilidad empresarial. Revista OIDLES 2008, 1, 1-25.
27. Carballo Penela, A.; García-Negro, M.C.; Doménech Quesada, J.L.; Villasante, C.S.; Rodríguez Rodríguez, G.; García Arenales, M. A pegada ecolóxica corporativa: concepto e aplicación a dúas empresas pesqueiras de Galicia. Revista Galega de Economía 2008, 17, 149-176.

28. Caselles Moncho, A.; Carrasco Esteve, M.; Martínez Gascón, A.; Coll Ribera, S.; Doménech, J.L.; González Arenales, M. La huella ecológica corporativa de los materiales: aplicación al sector comercial. Revista OIDLES 2008, 1, 1-24.

29. Coto Millán, P.; Mateo Mantecón, I.; Doménech, J.L.; Quesada, Y.; González-Arenales, M. La Huella Ecológica de las Autoridades Portuarias y los Servicios. Revista OIDLES 2008, 1, 1-27.

30. Doménech, J.L.; González-Arenales, M. La huella ecológica de las empresas: 4 años de seguimiento en el Puerto de Gijón. Revista OIDLES 2008, 1, 1-23.

31. Marañon, E.; Iregui, G.; Doménech, J.L.; Fernández-Nava, Y.; González-Arenales, M. Propuesta de índices de conversión para la obtención de la huella de los residuos y los vertidos. Revista OIDLES 2008, 1, 1-22.

32. IPCC (International Panel on Climate Change). Greenhouse Gas Inventory: workbook. Revised 1996 IPCC Guidelines, Vol 2. Available online: http://www.ipcc-nggip.iges.or.jp/public/gl/nrgspan.html (accessed October 6, 2008).

33. Wackernagel, M. The Ecological Footprint of Italia: Calculation Sheet. Available online: http://www.iclei.org/ICLEI/ef-ita.xls (accessed November 20, 2005).

34. Ibañez Etxeburúa, N. La huella ecológica de Donostia-San Sebastián. Available online: http://www.agenda21donostia.com/cas/corporativa/docs/huellaeco.pdf (accessed November 1, 2005).

35. Mayor Farguell, X.; Quintana Gozalo, V.; Belmonte Zamora, R. Aproximación a la huella ecológica de Cataluña. Available online: http://www.catsostenible.org/pdf/DdR_7_Huella_Ecologica.pdf (accessed Novomber 6, 2006).

36. Álvarez Díaz, P.D.; Doménech Quesada, J.L.; Perales Vargas-Machuca, J.A. Huella ecológica energética corporativa: Un indicador de la sostenibilidad empresarial. Revista OIDLES 2008, 1, 1-25.

37. Suh, S.; Lenzen, M.; Treloar, G.J.; Hondo, H.; Horvath, A.; Huppes, G.; Jolliet, O.; Klann, U.; Krewitt, W.; Moriguchi, Y.; Munksgaard, J.; Norris, G. System Boundary Selection in Life-cycle Inventories. Environ. Sci. Technol. 2004, 38, 657-664.

38. Aranda, A.; Zabalza, I.; Scarpellini, S. Economic and Environmental Analysis of the Wine Bottle Production in Spain by means of Life Cycle Assessment. International Journal of Agricultural Resources, Governance and Ecology (Special Issue on Life Cycle Assessment in the Terciary Sector) 2005, 4, 178-191.

39. Fullana, P.; Gazulla, C.; Clavijo, M.J.; Puerta, M.; Tubilleja, M. Análisis del ciclo de vida del vino de crianza D.O.C. Rioja. Dirección General de Calidad Ambiental, Consejería de Turismo, Medio Ambiente y Política Territorial del Gobierno de La Rioja, 2005.

Table 1 is not available in this version of the article. To view this additional information, please use the citation on the first page of this chapter.

CHAPTER 2

ENVIRONMENTAL IMPACTS OF CONSUMPTION OF AUSTRALIAN RED WINE IN THE UK

DAVID AMIENYO, CECIL CAMILLERI, AND ADISA AZAPAGI

2.1 INTRODUCTION

The global production level of wine stands at around 27 billion litres a year (Key Note, 2011). The UK consumes 12.9 million hectolitres1 or 4.8% of the world's wine production; this is equivalent to 21 l per capita per year (HMRC, 2012). However, only 3% of this is produced in the UK, so that the UK wine sector is heavily dependent on imports. It is thus not surprising that, with an estimated value of £11.8 billion (retail selling price), the UK was the world's largest market in 2010 for imported wines (Key Note, 2011). Prior to that, the UK was ranked 3rd behind the USA and France in terms of national shares of world wine consumption value (Anderson and Nelgen, 2011), while for total consumption volumes for 2011 the UK was ranked 6th behind France, USA, Italy, Germany and China (OIV, 2012). As shown in Fig. 1, Australian wines are most popular with the UK consumer, with around 17% adults buying these wines (Key Note, 2011). The next most popular are French wines (13%).

Environmental Impacts of Consumption of Australian Red Wine in the UK. © Amienyo D, Camilleri C, and Azapagi A. Journal of Cleaner Production, **72,**1 (2014). DOI: 10.1016/j.jclepro.2014.02.044. *Reprinted with permission from the authors.*

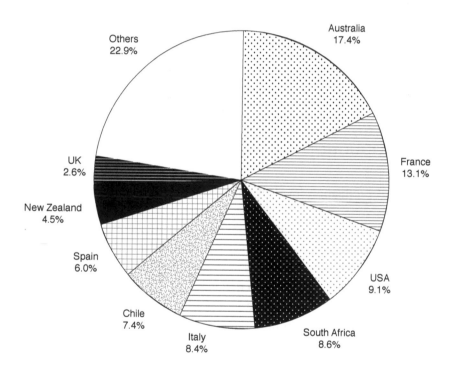

FIGURE 1: Wine purchased in the UK by country of origin (Data from Key Note, 2011). [Data represent percentage of adults buying wine. Others: Germany 3.8%; Argentina 1.7%; Portugal 1.3%; Bulgaria 0.5%; Other 3.8%].

The environmental impacts from wine consumption in the UK are un-known apart from few estimates. For example, it has been suggested that wine consumption contributes around 0.4% of the total UK greenhouse (GHG) emissions (Garnett, 2007) and 559,000 tonnes of packaging waste per year (Jenkin, 2010). On a global scale, the study by Rugani et al. (2013) estimates that the wine sector is responsible for around 0.3% of annual global GHG emissions from anthropogenic activities. An extensive body of literature exists on the environmental impacts of wines produced in various regions, including Canada (Point et al., 2012), Italy (Notarnicola et al., 2003, Ardente et al., 2006, Petti et al., 2006, CIV, 2008a, CIV, 2008b, Pizzigallo et al., 2008, Benedetto, 2010, Benedetto, 2013 and Cichelli et al., 2010); New Zealand (Herath et al., 2013), Portugal (Neto et al., 2013), Spain (Aranda et al., 2005, Panela et al., 2009, Gazulla et al., 2010, Vázquez-Rowe et al., 2012a, Vázquez-Rowe et al., 2012b and Villanueva-Rey et al., 2014) and the USA (Colman and Päster, 2007). Most studies have focused on GHG emissions and their review can be found in Rugani et al. (2013). Studies that have considered other environmental impacts in addition to the GHG emissions include Notarnicola et al., 2003, Aranda et al., 2005, CIV, 2008a and CIV, 2008b), Gazulla et al. (2010) and Point et al. (2012).

This paper sets out to estimate the life-cycle environmental impacts of red wine produced in Australia and consumed in the UK. In the first part of the paper, the focus is on the wine produced by one of largest producers in South Australia, which itself is the largest wine-producing state in the country (Fearne et al., 2009). These results are then used to estimate the environmental impacts at the sectoral level related to the consumption of Australian red wine in the UK. Several options for reducing the impacts are also considered, including transport of bulk rather than bottled wine, increased recycling and light-weighting of glass bottles, as well as use of carton packaging.

2.2 METHODS

2.2.1 GOAL AND SCOPE OF THE STUDY

The main goal of this study is to estimate the environmental impacts and identify improvement opportunities in the life cycle of red wine produced in Australia and consumed in the UK. The analysis is carried out in two stages. First, the environmental impacts are calculated based on the functional unit defined as 'production and consumption of a 0.75 l bottle of wine'. In the second stage, the functional unit is the 'annual consumption of Australian red wine in the UK' to determine the total impacts from its consumption.

As shown in Fig. 2, the system boundary of the study is from 'cradle to grave', comprising the following stages, described in more detail in subsequent sections:

- Viticulture: water supply, production of fertilisers and pesticides and fuels for cultivation and harvest of wine grapes;
- Packaging: production of primary packaging comprising glass bottles, cork stoppers and paper labels;
- Vinification and bottling: production and bottling of wine, generation and consumption of electricity; production and consumption of auxiliary materials including water, sulphur dioxide and sodium hydroxide used in the production of wine;
- Transport: transport of packaging materials to the winery, bottled wine to UK retailers and post-consumer waste packaging to waste management; and
- Waste management: wastewater treatment of effluents from the winery and disposal of in-process and post-consumer wastes.

The following activities are outside the system boundary of the study for the following reasons:

- Secondary and tertiary packaging for the wine: their contribution to the impacts is assumed to be small based on the finding that it accounts for less than 1% of the total carbon footprint of wine (BIER, 2012); this exclusion is also common in other studies (e.g. Gazulla et al., 2010 and Point et al., 2012).
- Yeasts, filtering and clarifying agents, bacteria, enzymes and antioxidants used in the manufacturing process: the study by Notarnicola et al. (2003) estimates that the contribution of these auxiliary materials to the overall impacts is negligible.
- Transport of consumers to purchase the wine: transport of consumers to and from the point of retail purchase is not considered owing to a large uncertainty related to consumer behaviour and related allocation of impacts to a bottle of wine relative to other items purchased at the same time; this also is congruent with the PAS 2050 standard (BSI, 2011).

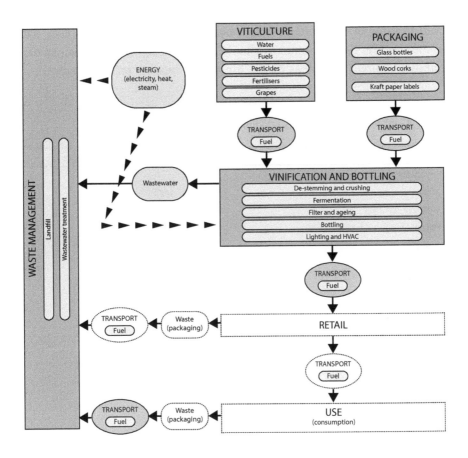

FIGURE 2: The life cycle of wine [—Stages excluded from the system boundary. HVAC: heating, ventilation and air conditioning].

2.2.2 INVENTORY DATA AND ASSUMPTIONS

Primary production data have been obtained from a wine producer, including the amount of fuels, fertilisers and pesticides for viticulture as well as electricity and auxiliary materials used for vinification. All other data have been sourced from Ecoinvent (2010), Gabi (PE, 2011) and CCaLC (2013). Data from open literature have also been used to estimate inventory data where specific data were not available. More detail on the inventory data and their sources as well as on the life-cycle stages is provided below.

TABLE 1: Inventory data for grape viticulture, vinification and wine bottling[a].

Inputs	Amount per 0.75 l of wine
Viticulture	
Water	362 l
Nitrogen fertiliser	9.6 g
Phosphorus fertiliser	27.8 g
Pesticides	9.8 g
Electricity	37 Wh
Diesel	0.074 l
Petrol	0.032 l
Vinification and bottling	
Wine grapes	1.05 kg
Water	1.31 l
Electricity	115 Wh
Glass bottles (0.75 l; 85% recycled glass content)	465 g
Cork stoppers	5.25 g
Kraft paper labels	1.05 g
VOC emissions from fermentation[b]	0.4 g

[a] All life-cycle inventory data are from the Ecoinvent database (2010) except for the data for glass bottles which are from CCaLC (2013). [b] Volatile Organic Compounds (VOC) as ethanol. Source: US EPA (1995). Source: US EPA (1995).

2.2.2.1 VITICULTURE

Conventional cultivation of the Shiraz grape, the predominant viticulture practice and grape varietal in South Australia (Wine Australia, 2013), has been assumed in this study. The average grape yield has been estimated at just over 10 t/ha (Anderson and Nelgen, 2011). This falls within the range of 6–12 t/ha.year estimated by Notarnicola et al. (2003) for other regions.

As shown in Table 1, the main input materials for cultivation of the grapes are irrigation water, fertilisers and pesticides. Additionally, diesel and petrol are used for the agricultural machinery. Note that site-specific dispersion of nutrients and pesticides has not been considered owing to a lack of site-specific dispersion models to estimate the fate of these emissions to the air, water and soil (Milà i Canals and Polo, 2003). This is consistent with some other studies (e.g. Point et al., 2012) but it may have an effect on eutrophication and the toxicity-related impacts from the viticulture stage and should be borne in mind when interpreting the results.

2.2.2.2 VINIFICATION AND BOTTLING

To produce 0.75 l of red wine, 1.05 kg of grapes are required (Table 1). This is also within the range estimated by some other authors (e.g. Notarnicola et al., 2003, Benedetto, 2013 and Villanueva-Rey et al., 2014). The wine production process begins with de-stemming and crushing of the grapes to obtain a liquid must (juice). Prior to fermentation, the temperature of the must can be adjusted to allow a period of cold maceration during which the grape skins are softened by soaking. This allows the extraction of tannins and flavour compounds into the must from the grape skins which is crucial to red-wine making. This is also the process during which red wine gets its colour as most musts are clear or greyish in colour. The must is then fermented at a temperature between 28 and 30 °C, during which yeast is fed into the fermentation vessel to convert the sugars into alcohol. After fermentation, which usually lasts between one and two weeks, the wine is pumped off into tanks and the skins are pressed to extract the remaining juice and wine. Secondary fermentation is then carried out using bacteria to decrease the acidity and soften the taste of wine by converting malic to lactic acid. Prior to bottling, the wine must be settled, clarified and finally filtered. Most red wine is then matured in oak barrels for a few weeks to several years depending on the variety of grapes and desired wine style (Wine Australia, 2012).

The inventory for vinification and bottling is based on primary production data provided by the wine producer. Life cycle impacts of electricity generation have been modelled based on the relative shares of primary energy resources in the Australian grid. Fugitive emissions of volatile organic compounds (VOC) from fermentation have also been considered (see Table 1). The co-products from wine-making (pomace, lees and press syrup) have been assumed to be disposed of as waste owing to a negligible economic value. A similar approach has also been taken by some other authors (see Rugani et al., 2013).

2.2.2.3 PACKAGING

The wine is packaged in 0.75 l bottles using cork stoppers and paper labels (Table 1). Data on the components and weights of the bottles have been obtained from Gujba and Azapagic (2011) and CCaLC (2013). The life cycle impacts for manufacturing and recycling of glass bottles have been modelled based on industry-specific data available in CCaLC (2013). The bottles are assumed to contain 85% recycled content and the emissions associated with the recycling process have been allocated to the life-cycle stage that utilises the recycled material, thereby displacing a portion of virgin material. This approach has also been adopted by the beverage industry in studies of the carbon footprint of various beverages, including wine (BIER, 2012). Different percentages of recycled glass content are also considered in the sensitivity analysis to examine the effect of this parameter on the environmental impacts.

2.2.2.4 TRANSPORT

After bottling, the wine is shipped to the UK (Table 2). Shipping bottled rather than the wine in bulk is the usual practice for wines imported to the

UK from Australia (Garnett, 2007). Transport distances have been estimated based on the data obtained from the wine manufacturer. Owing to a lack of specific data, the trucks are assumed to have a total capacity of 40 tonnes. Generic distances of 50 km have been used for transport of wine grapes, cork stoppers and paper labels as well as post-consumer waste, for which specific data have not been available.

TABLE 2: Transport modes and distances in the wine supply chain[a].

Material	Transport mode	Distance (km)
Wine grapes	Truck (40 t)	50
Glass bottles	Truck (40 t)	39
Cork stoppers	Truck (40 t)	50
Kraft paper labels	Truck (40 t)	50
Bottled wine	Truck (40 t)	128
	Container ship	20,030
Post-consumer packaging waste	Truck (40 t)	50

[a]*All life cycle inventory data are from the Gabi database (PE, 2010), except for the data for the container ship which are from ILCD (2010).*

2.2.2.5 WASTE MANAGEMENT

As indicated in Table 3, the waste streams considered are effluents from the winery and post-consumer waste packaging; these data have been obtained from the wine manufacturer. Given that the wine is consumed and bottles discarded in the UK, the UK waste management practice is assumed for waste bottles, with 85% of glass recycled and the rest landfilled together with the labels; all post-consumer cork is also landfilled (WRAP, 2007 and British Glass, 2007). As mentioned earlier, the effect on the results of different glass recycling rates is considered as part of the sensitivity analysis later in the paper.

TABLE 3: Waste management[a].

Waste	Amount (g/0.75 l of wine)	Waste management option
Effluents from the winery	615	Wastewater treatment
Glass	465	85% recycled, 15% landfilled
Paper label	1.05	Landfilled
Wood cork	5.25	Landfilled

[a]*All life cycle inventory data are from the Gabi database (PE, 2010).*

2.3 RESULTS AND DISCUSSION

The Gabi 4.3 LCA software has been used to model the system and the CML 2001 impact assessment method (Guinee et al., 2001) has been used to calculate the environmental impacts. The CML method has been selected because of its coverage of a wide range of environmental impacts relevant to wine and the regions covered by the study. As a mid-point method, it also helps to preserve transparency by allowing an analysis of individual impacts rather than aggregating them in to a single measure of environmental 'damage' as is the case in end-point approaches such as Ecoindicator 99 (Goedkoop and Spriensma, 2001). In addition to the CML impact categories, the primary energy consumption and water demand have also been estimated. The results are first presented for a bottle of wine, followed by the discussion of the environmental implications associated with the annual consumption of Australian red wine in the UK.

2.3.1 ENVIRONMENTAL IMPACTS OF A BOTTLE OF WINE

The life-cycle environmental impacts of wine are shown in Fig. 3. For example, a bottle of wine requires around 21 MJ of primary energy, 363 l of water and contributes 1.25 kg CO_2 eq. to the global warming potential (GWP).As indicated in Fig. 3, viticulture, transport and packaging are the major contributors to most of the impacts. The former is the hot spot for eight out of 12 categories considered: primary energy demand (PED), water demand (WD), abiotic depletion (ADP), GWP, human toxicity (HTP), and marine, freshwater and terrestrial ecotoxicity (MAETP, FAETP and TETP, respectively). The results suggest that emissions arising from the production and use of pesticides, fertilisers and fuels are the main contributors to the impacts from viticulture. For the example of GWP, pesticides and fertilisers collectively contribute 82% and fuels the remaining 18%.

Transport is the key contributor to the acidification (AP), eutrophication (EP), GWP, ozone depletion (ODP) and photochemical oxidant creation (POCP) potentials. The majority of these impacts are due to wine shipping. For instance, shipping is responsible for 84% of GWP from transport, with the remaining 16% being from road transport, in both cases owing to the emissions of CO_2. The other impacts from shipping are mainly due to the emissions of SO_2 and NO_x. Packaging is an important contributor to PED, AD, GWP, HTP, MAETP, FAETP and ODP, accounting for over 20% to each impact and, in the case of HTP and MAETP, over 40%. Emissions of selenium in the production of glass bottles are largely responsible for HTP and hydrogen fluoride for MAETP.

The vinification stage is a hot spot for EP, accounting for 30% of the total, largely because of the emissions of organic compounds to freshwater arising from the winery effluents. For all other impacts, its contribution is on average 10% or less. The contribution of fugitive VOC emissions from grape fermentation to GWP and POCP is negligible (0.03% and 0.004%, respectively) as is that of waste management (≤1%).

2.3.1.1 COMPARISON OF RESULTS WITH OTHER STUDIES

As mentioned in the introduction, a number of studies of the environmental impacts of wine have been carried out, but comparison of the results is difficult because of different geographical regions, varying agricultural practices as well as study assumptions and data sources. Nevertheless, an attempt is made here to compare the results of this with some other studies. As the GWP has been studied most extensively, these results are discussed first, followed by a comparison of other impacts (for studies where they have been estimated and comparison has been possible).

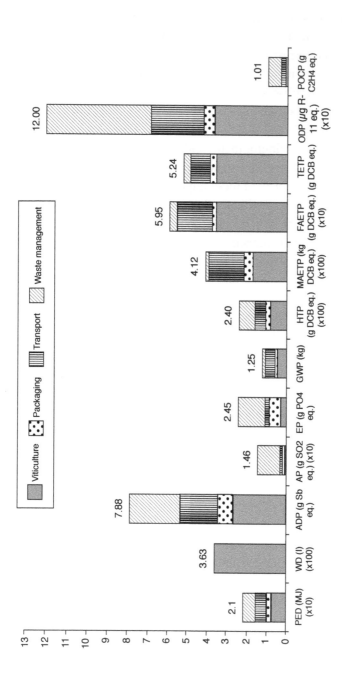

FIGURE 4: Comparison of environmental impacts of wine obtained in this and other studies (functional unit: 0.75 l of wine) [aLife cycle impacts of conventional wine (different varieties and styles) produced and consumed in Canada, minus consumer shopping and storage. bLife cycle impacts of conventional red wine produced in Spain and transported to the UK, minus impacts from secondary packaging (barrel production). For full names of the impact categories, see caption for Fig. 3. The scaled values should be multiplied with the factor shown in brackets against the relevant impact to obtain the original values. Note the comparison of other environmental impacts is not possible as the other studies did not consider them.].

The review study by Rugani et al. (2013) provides estimates of the GWP of red wine obtained in around 30 LCA studies worldwide. The authors observe a large variation in the results, ranging from 0.83 to 3.51 kg CO_2 eq. per bottle, with an average estimated at 2.17 kg CO_2 eq. Therefore, the total GWP obtained in the current study falls well within the reported range. The results are also within the range for the cradle-to-gate GWP, here estimated at 0.86 kg CO_2 eq. per bottle; this compares to 0.26–1.92 kg CO_2 eq. reported in literature (see Rugani et al., 2013).

The results for the other environmental impacts are compared to those reported by Gazulla et al. (2010) and Point et al. (2012) in Fig. 4. Comparison with the other studies is not possible either because they do not include impacts other than GWP or because of different assumptions. As can be seen from the figure, there are significant differences between the results in the three studies owing to different geographical regions, waste management options, bottle weights and recycled content as well as wine distribution scenarios. The closest agreement is found between the current and Point et al. study for the ODP while the largest difference is for the POCP which is around 70% higher in this research than in either of the two studies, largely because of the long-distance transport from Australia to the UK.

2.3.2 ENVIRONMENTAL IMPACTS OF CONSUMPTION OF AUSTRALIAN RED WINE IN THE UK

The environmental impacts from the consumption of Australian red wine in the UK have been estimated by scaling up the LCA results for a bottle of wine to the annual consumption of the wine obtained through market analysis. As mentioned before, the Australian wine considered in this study can be assumed to be representative of Australian red wines in general for several reasons. First, South Australia represents the largest wine making region in Australia (WRAP, 2007 and Fearne et al., 2009) and this study considers a producer in this region. Secondly, the data are sourced from one of the ten largest producers by sales of branded wine in Australia (Winebiz, 2013). However, it should be noted that all sectoral assessments carry some uncertainty owing to limited data availability and the need to extrapolate the results. Nevertheless, this kind of analysis helps to put environmental impacts into broader, national context and identify opportunities for improvements at the sectoral level.

In 2009, a total of 2.2 million hectolitres of Australian wine were imported into the UK (New Zealand Trade and Enterprise, 2011) of which 57% was red and 40% white wine with the rest being sparkling wine (PIRSA, 2010). Extrapolating the above LCA results for the amount of red wine, gives the total environmental impacts for 1.25 million hl of red wine as shown in Fig. 5. For example, the total annual PED is estimated at around 3.5 PJ, water consumption is 600 million hl and the GWP is around 210,000 tonnes CO_2 eq. The latter represents about 0.08% of the consumption-based GHG emissions from total annual UK imports in 2011 (Defra, 2013)3. To put these results further into context, assuming that the average GWP of all wine is 2.2 kg/bottle (Rugani et al., 2013), then 12.9 million hectolitres of wine consumed in the UK annually (HMRC, 2012) emit 3.78 million tonnes of CO_2 eq. per year or 0.6% of the UK emissions.3 This is slightly higher than the previously mentioned estimate of 0.4% (Garnett, 2007). While this percentage appears to be small, it is equivalent to the annual GHG emissions from 1 million cars. By comparison, the estimated annual contribution of wine to the global GHG emissions is 0.3% (Rugani et al., 2013). Putting the other environmental impacts in context is more difficult owing to a lack of data at the national level.

The following sections explore how the impacts from the consumption of Australian wine in the UK could be reduced. Given the high contribution of transport and packaging, opportunities for improvements in these two life-cycle stages are examined. Although the considerations here refer to Australian wine, the reduction strategies considered also apply to wines from other regions.

Note that, although viticulture is also an environmental hot spot owing to the use of fertilisers and pesticides, improvements in stage are not considered. The reason for this is that fertiliser and pesticide inputs are based on the actual data obtained from the wine producer, representing a viticulture practice optimised over the years. Therefore, there is little scope for improvement in this stage.

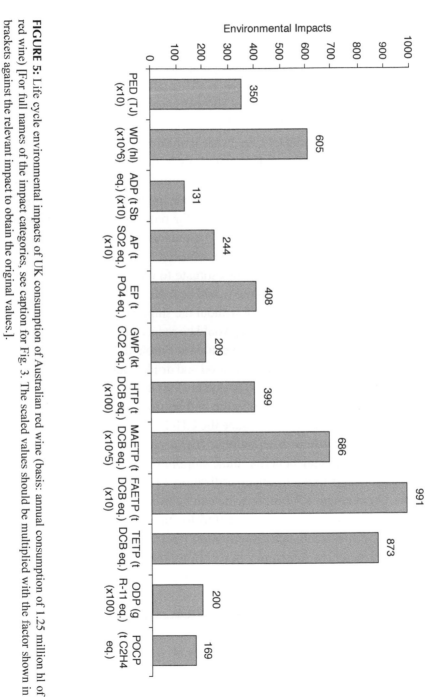

FIGURE 5: Life cycle environmental impacts of UK consumption of Australian red wine (basis: annual consumption of 1.25 million hl of red wine) [For full names of the impact categories, see caption for Fig. 3. The scaled values should be multiplied with the factor shown in brackets against the relevant impact to obtain the original values.].

2.3.2.1 IMPROVEMENT OPPORTUNITIES FOR SHIPPING

This analysis focuses on GWP owing to a lack of data on the effect of shipping on other environmental impacts. This study indicates that shipping bottled wine from Australia to the UK accounts for 84% of the GWP from transport and 26% of the total GWP (see Section 3.1). This means that shipping adds around 0.33 kg CO_2 eq./bottle to the total GWP from wine. As this is a significant contribution, found not only in this but other studies related to wine shipping (e.g. WRAP, 2007), it is important to look at alternative options. For example, bulk shipping of wine and bottling closer to the consumer has been suggested as an option for reducing the GHG emissions from wine. The study by WRAP (2007) estimates that a GWP saving of 0.16 kg CO_2 eq. per bottle of wine can be achieved by shipping it in bulk from Australia to the UK. Applying this estimate to the results from the current study indicates that the total GWP would be reduced by 13% to 1.09 kg CO_2 eq./bottle. For context, the GWP without the shipping from Australia (i.e. if the wine was produced in the UK) would be reduced by 26% to 0.93 kg CO_2 eq./bottle. Annually, bulk shipping could save around 27,000 t CO_2 eq. This would contribute 0.4% towards the food and drinks industry's aim to reduce its CO_2 emissions by 35% by 2020 on the 1990 levels (FDF, 2012).

Therefore, bulk shipping of wine and bottling closer to the final market should be encouraged to reduce the GHG and other emissions, particularly SOx (because of high-sulphur fuel used in shipping). In addition to the environmental benefits, bulk shipping would also result in cost savings through more efficient utilisation of container space (WRAP, 2008). On average, 67% more wine can be transported by shipping wine in flexi-tanks or ISO tanks, compared to standard container shipping.5 However, issues such as contamination (from residues of previous cargoes) and negative consumer perceptions may hinder the widespread adoption of bulk shipping of wine (WRAP, 2008).

2.3.2.2 IMPROVEMENT OPPORTUNITIES FOR PACKAGING

Glass bottles contribute over 40% to the HTP and MAETP and 20% to the other environmental impacts (see Fig. 3). Thus, the sections below explore the effect of two parameters on the environmental impacts from wine: recycled glass content and bottle light-weighting. A further packaging option is also considered: using carton containers instead of glass bottles, as practiced by some wine producers (FDIN, 2012).

2.3.2.2.1 RECYCLED GLASS CONTENT

To examine the effect of glass recycling on the environmental impacts, a range from 0% to 100% recycled glass content has been considered. The results are shown in Fig. 6 for the annual wine consumption (the trend is the same per bottle of wine and hence not shown). For example, for each 10% increase in the amount of recycled glass, GWP is reduced by 2%. This amounts to a saving of 22 g CO_2 eq. per bottle of wine or around 3600 tonnes annually. The savings are due to lower energy consumption for bottle manufacturing and reduced amount of post-consumer waste being landfilled. The savings for the other environmental impacts are smaller and range from 0.7% (POCP) to 1.2% (HTP). Vázquez-Rowe et al. (2012a) also observed similar environmental savings from recycling of glass bottles. Although the savings appear relatively small per bottle of wine, they are nevertheless significant at the sectoral level. Thus, there is a clear case for increasing the recycled content of glass packaging to reduce the environmental impacts from wine.

2.3.2.2.2 BOTTLE LIGHT-WEIGHTING

The results in Fig. 7 show that reducing the weight of glass bottles by 10% results in GWP savings of about 4% or 43 g CO_2 eq. per bottle; this is equivalent to around 7000 tonnes of CO_2 eq. based on the amount of Australian red wine consumed per year. Savings in other impacts range from 3% (TETP) to 7% (ODP). For 30% lighter bottles, the GWP would be reduced by 11% with the other impacts decreasing by 7–15%. These reductions are due to lower energy and material consumption for manufacturing of glass bottles and reduced impacts from transporting less glass. Similar savings were estimated in the study by Point et al. (2012) for wine in Canada who found that 30% lighter bottles saved from 2% to 10% of the impacts. Thus, these results indicate that light-weighting is an important option for reducing the environmental impacts in the wine sector.

2.3.2.2.3 CARTONS VS. GLASS BOTTLES

In addition to increasing the recycled content and light-weighting of glass bottles, the effect of using carton packaging instead of glass bottles has also been assessed. It has not been possible to ascertain the volume of wine packaged in cartons in the UK market owing to a lack of data. However, recent reports show that a number of wine producers and importers were starting to introduce wines packaged in cartons into the UK market (FDIN, 2012). Currently, around 10% of still wines in the global market are packaged in cartons (FDIN, 2012).

As can be seen in Fig. 8, packaging wine in cartons instead of bottles results in savings in all the environmental categories considered, except for water demand which is close to the glass bottle system (however the latter should be interpreted with care owing to a general lack of data on water consumption in LCA databases). For example, compared to the current operations, packaging wine in cartons reduces the GWP by 51%, from 1.25 to 0.62 kg CO_2 eq. per bottle of wine compared to the glass bottles. For the other environmental impacts, the savings range from 25% (EP) to 70% (MAETP). The savings arise mainly from reduction in the energy required for manufacturing of the packaging and reduced emissions from transporting the significantly lighter cargo.

Scaling the results up for the annual consumption of Australian red wine and assuming (hypothetically) that 10% of the Australian red wine consumed in the UK is packaged in cartons, the annual savings of GHG emissions would be around 5% or 10,600 tonnes CO_2 eq. (Fig. 9). Savings for the other environmental impact categories range from 2.5% (EP) to 7% (MAETP). These results show that, on the environmental basis, there is a compelling case for a widespread adoption of cartons in the wine industry. However, other factors such as economic aspects, consumer perception, ease of transportation, shelf life and potential impacts on the glass-bottle industry need to be investigated to understand the full sustainability impacts of the wider use of carton packaging for wine.

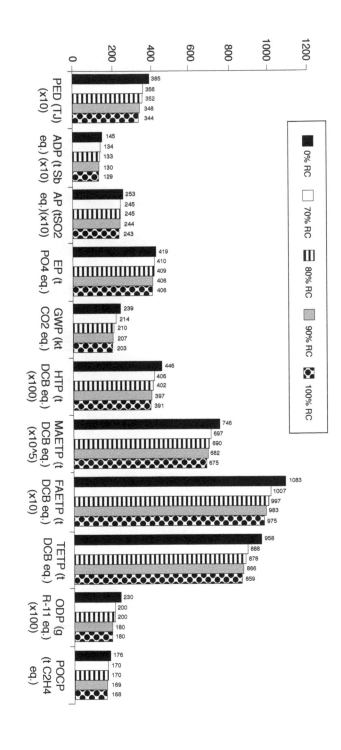

FIGURE 6: Effect of recycled glass content on the environmental impacts of wine (basis: annual consumption of 1.25 million hl of red wine) [Water demand not shown as it is not affected by recycled glass content. For full names of the impact categories, see caption for Fig. 3. The scaled values should be multiplied with the factor shown in brackets against the relevant impact to obtain the original values.].

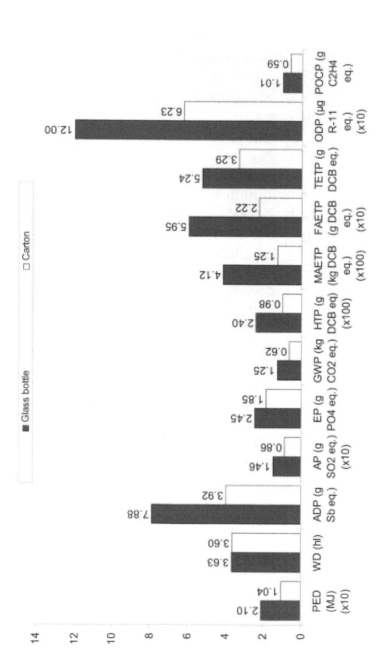

FIGURE 7: Effect of light-weighting on the environmental impacts of wine (basis: annual consumption of 1.25 million hl of red wine) [For full names of the impact categories, see caption for Fig. 3. The scaled values should be multiplied with the factor shown in brackets against the relevant impact to obtain the original values.].

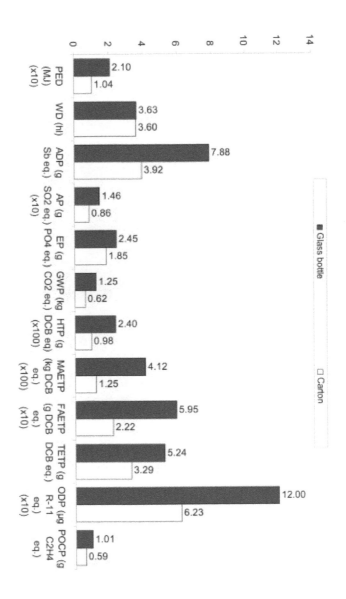

FIGURE 8: Effect of using carton packaging on the environmental impacts of wine (functional unit: 0.75 l of wine) [Glass bottle: 85% recycled glass content, end-of-life waste management as in Table 3. Carton: 100% virgin component materials (cardboard, plastic film and aluminium foil), end-of-life waste management: 100% landfill. Data for carton packaging from CCaLC (2013). For full names of the impact categories, see caption for Fig. 3. The scaled values should be multiplied with the factor shown in brackets against the relevant impact to obtain the original values.].

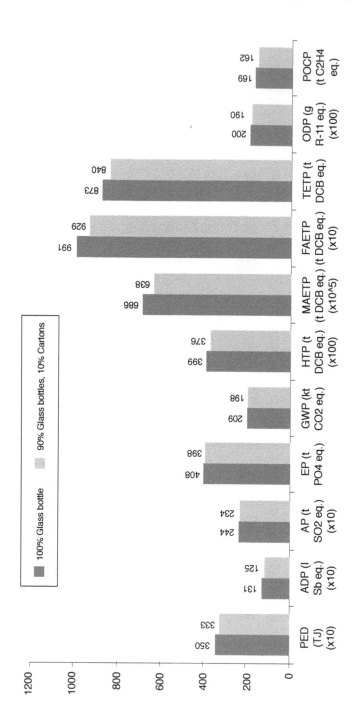

FIGURE 9: Effect of using carton packaging on the environmental impacts of wine (basis: 1.25 million hl of red wine) [For assumptions see the caption for Fig. 8. For full names of the impact categories, see caption for Fig. 3. The scaled values should be multiplied with the factor shown in brackets against the relevant impact to obtain the original values.].

2.4 CONCLUSIONS

This paper has presented and discussed the life-cycle environmental impacts of consumption of Australian red wine in the UK. The results have been estimated first for one bottle and then for the total annual consumption of Australian red wine in the UK. For example, it has been found that a bottle of wine requires 21 MJ of primary energy, 363 l of water and generates 1.25 kg of CO_2 eq. Extrapolating these results to the annual UK consumption of Australian red wine gives the total primary energy demand of around 3.5 PJ, water consumption of around 600 million hl and GWP of 210,000 tonnes CO_2 eq. The latter represents about 0.08% of the consumption-based GHG emissions from total annual UK imports in 2011. A total of 12.9 million hectolitres of wine consumed in the UK annually emits 2.8 million tonnes of CO_2 eq. per year or 0.6% of total the UK emissions.

The results show that viticulture is the main hot spot in the life cycle of wine, contributing on average 41% to the impacts; this is mainly due to the life cycles of pesticides, fertilisers and fuels. Transport is the next largest contributor adding on average 32% to the impacts, largely from the shipping of wine to the UK from Australia. For instance, shipping generates around 0.33 kg CO_2 eq. per bottle of wine. The impacts of packaging are also significant, contributing on average 24%, mainly owing to the production of glass bottles. Finally, vinification contributes around 8% while the impacts from the end-of-life management are small (1%).

Several options for reducing the impacts from wine have been considered based on the identified hot spots: shipping of bulk rather than bottled wine, increased recycling and light-weighting of glass bottles as well as using carton packaging instead of bottles. The results suggest that bulk shipping would reduce the GWP by 13% to 1.09 kg CO_2 eq./bottle, saving 27,000 t CO_2 eq. annually. This would contribute 0.4% towards the food and drinks industry's aim to reduce its CO_2 emissions by 35% by 2020 on the 1990 levels.

It has also been found that for every 10% increase in the amount of recycled glass, the GWP is reduced by about 2%. This amounts to a saving of 22 g CO_2 eq./bottle or around 3600 tonnes per year. The savings are transportation and economic impacts on the glass bottle industry as well as consumer perception would need to be investigated further. The latter two also apply to packing the wine into cartons. Further aspects, including economic and health costs and benefits should also be considered in future work to gain a better understanding of the full life-cycle sustainability implications of wine production and consumption.

REFERENCES

1. K. Anderson, S. Nelgen. Global Wine Markets, 1961 to 2009: a Statistical Compendium. University of Adelaide Press, Adelaide (2011)
2. A. Aranda, I. Zabalza, S. Scarpellini. Economic and environmental analysis of wine bottle production in Spain by means of life cycle assessment. Int. J. Agric. Resour. Gov. Ecol., 4 (2) (2005), pp. 178–191
3. F. Ardente, G. Beccali, M. Cellura, A. Marvuglia. POEMS: a case study of an Italian wine-producing firm. Environ. Manag., 38 (2006), pp. 350–364
4. G. Benedetto. Life cycle environmental impact of Sardinian wine. Paper Prepared for the 119th EAAE Seminar 'Sustainability in the Food Sector: Rethinking the Relationship between the Agro-food System and the Natural, Social, Economic and Institutional Environments', Capri, Italy, June 30th – July 2nd (2010) www.centroportici.unina.it/EAAE_Capri/papers/Session11/Benedetto.pdf
5. G. Benedetto. The environmental impact of a Sardinian wine by partial life cycle assessment. Wine Econ. Policy, 2 (2013), pp. 33–41
6. BIER. Research on the Carbon Footprint of Wine. Beverage Industry Environmental Roundtable (2012) www.bieroundtable.com/files/Wine%20Final%20DEP.pdf
7. British Glass. Sustainability Report 2007. (2007) www.britglass.org.uk/Industry/Recycling.html
8. BSI. Publicly Available Specification PAS 2050:2008. Specification for the Assessment of the Life Cycle Greenhouse Gas Emissions of Goods and Services British Standards Institution, London (2011)
9. CCaLC. CCaLC Tool and Database V3. (2013) www.ccalc.org.uk
10. A. Cichelli, A. Raggi, C. Pattara. Life Cycle Assessment and Carbon Footprint in the Wine Supply-chain. (2010) www.ecososteniblewine.com/files/P17_Cichelli_LCA_Carbon_Footprint.pdf
11. CIV. Environmental Product Declaration: Bottled Red Sparkling Wine "Grasparossa Righi". CIV (2008) March 2008 http://gryphon.environdec.com/data/files/6/7505/EPD%20S-P-00109%20ingl-2008-def.pdf
12. CIV. Environmental Product Declaration: Climate Declaration for Bottled Organic Lambrusco Grasparossa Red Sparkling Wine "Fratello Sole". CIV (2008) Zeko5m@Zeko5m@March 2008 http://gryphon.environdec.com/data/files/6/7521/EPD_Fratello%20Sole_english%202008-def.pdf

13. T. Colman, P. Päster. Red, White and "Green": the Cost of Carbon in the Global Wine Trade. AAWE Working Paper No. 9. October 2007 American Association of Wine Economists (2007) www.wine-economics.org/workingpapers/AAWE_WP09. pdf

14. Defra. Greenhouse Gas Conversion Factors for Company Reporting. 2012 Guidelines. (30 May 2012) www.defra.gov.uk

15. Defra. UK's Carbon Footprint 1997–2011. Department for Environment, Food and Rural Affairs, London (2013) www.defra.gov.uk

16. Ecoinvent Centre. Ecoinvent v2.2 Database. Swiss Centre for Life Cycle Inventories, Dübendorf, Switzerland (2010) www.ecoinvent.ch

17. FDF. Cutting CO2 Emissions. Food and Drink Federation, UK (2012) www.fdf.org. uk/priorities_sus_comp.aspx

18. A. Fearne, C. Soosay, R. Stringer, W. Umberger, B. Dent, C. Camilleri, D. Henderson, A. Mugford. Sustainable Value Chain Analysis: a Case Study of South Australian Wine. (January 2009) www.pir.sa.gov.au/data/assets/pdf_file/0003/93225/V2D_Final_Report.pdf

19. FDIN. First Carton Packaged Wines Now Available in the UK. (2012) www.fdin. org.uk/2012/09/first-carton-packaged-wines-now-available-in-the-uk/

20. T. Garnett. The Alcohol We Drink and its Contribution to the UK's Greenhouse Gas Emissions. Food Climate Research Network (2007) www.fcrn.org.uk/sites/default/ files/ALCOHOL%20final%20version%20TG%20feb%202007.pdf

21. C. Gazulla, M. Raugei, P. Fullana-i-Palmer. Taking a life cycle look at Crianza wine production in Spain: where are the bottlenecks?. Int. J. LCA, 15 (2010), pp. 330–337

22. M. Goedkoop, R. Spriensma. The Eco-indicator 99: a Damage Oriented Method for Life Cycle Assessment, Methodology Report. (third ed.)Pré Consultants, Amersfoort, The Netherlands (22 June 2001)

23. J.B. Guinée, M. Gorrée, R. Heijungs, G. Huppes, R. Kleijn, L. van Oers, A. Wegener Sleeswijk, S. Suh, H.A. Udo de Haes, H. de Bruijn, R. van Duin, M.A.J. Huijbregts. Life Cycle Assessment, an Operational Guide to the ISO Standards. Kluwer Academic Publishers, Dordrecht, The Netherlands (2001)

24. H. Gujba, A. Azapagic. Carbon footprint of beverage packaging in the United Kingdom. Chapter 37 M. Finkbeiner (Ed.), Towards Life Cycle Sustainability Management, Springer (2011), pp. 381–390 pp.591

25. I. Herath, S. Green, R. Singh, D. Horne, S. van der Zijpp, B. Clothier. Water footprinting of agricultural products: a hydrological assessment for the water footprint of New Zealand's wines J. Clean. Prod., 41 (2013), pp. 232–243

26. HMRC. Alcohol Bulletin July 2012 (31 August 2012) www.uktradeinfo.com/statistics/pages/taxanddutybulletins.aspx

27. ILCD. International Life Cycle Database, European Joint Research Centre (2010) http://lca.jrc.ec.europa.eu/lcainfohub/datasetArea.vm

28. N. Jenkin. Wine & Climate Change: Packaging & Transport Retail Programme, WRAP (2010) www.climatechangeandwine.com/conferencias/conf8/8_1.pdf

29. Key Note. Market Report Plus 2011: Wine Key Note Ltd, Middlesex (2011) www. keynote.co.uk/market-intelligence/view/product/10466/wine?medium=download

30. L. Milà i Canals, G.C. Polo. Life cycle assessment of fruit production. B. Mattson, U. Sonesson (Eds.), Environmentally-friendly Food Processing, Woodhead Publishing Ltd., Cambridge, England (2003), pp. 29–53

31. B. Neto, A. Dias, M. Machado. Life cycle assessment of the supply chain of a Portuguese wine: from viticulture to distribution. Int. J. Life Cycle Assess., 18 (2013), pp. 590–602

32. New Zealand Trade and Enterprise. Exporter Guide Wine in the United Kingdom and Ireland Market Profile (November 2011) www.nzte.govt.nz/explore-export-markets/market-research-by-industry/Food-and-beverage/Documents/Wine%20in%20UK%20and%20Ireland%20Nov%202011.pdf

33. B. Notarnicola, G. Tassielli, G. Nicoletti. Life cycle assessment (LCA) of wine production. B. Mattson, U. Sonesson (Eds.), Environmentally-friendly Food Processing, Woodhead Publishing Ltd., Cambridge, England (2003), pp. 306–325

34. OIV. Statistical Report on World Vitiviniculture 2012 International Organisation of Vine and Wine (2012) www.oiv.int/oiv/files/0%20-%20Actualites/EN/Report.pdf

35. A.C. Panela, Md.C. García-Negro, J.L.D. Quesada. A methodological proposal for corporate carbon footprint and its application to a wine-producing company in Galicia, Spain. Sustainability, 1 (2009), pp. 302–318

36. PE International. Gabi 4.3 (2010) www.gabi-software.com/databases/professional

37. L. Petti, A. Raggi, C. De Camilis, P. Matteucci, B. Sára, G. Pagliuca. Life cycle approach in an organic wine-making firm: an Italian case study. Proceedings Fifth Australian Conference on Life Cycle Assessment, Melbourne Australia, November 22nd–24th, 2006 (2006) www.conference.alcas.asn.au/2006/Petti%20et%20al.pdf

38. PIRSA. Wine Markets and Consumers Opportunities and Challenges for the Langhorne Creek Wine Region Primary Industries and Resources South Australia. (December 2010) www.pir.sa.gov.au

39. A.C.I. Pizzigallo, C. Granai, S. Borsa. The joint use of LCA and emergy evaluation for the analysis of two Italian wine farms. J. Environ. Manag., 86 (2008), pp. 396–406

40. E. Point, P. Tyedmers, C. Naugler. Life cycle environmental impacts of wine production and consumption in Nova Scotia, Canada. J. Clean. Prod., 27 (2012), pp. 11–20

41. B. Rugani, I. Vázquez-Rowe, G. Benedetto, E. Benedetto. A comprehensive review of carbon footprint analysis as an extended environmental indicator in the wine sector. J. Clean. Prod., 54 (2013), pp. 61–77

42. US EPA. Emission Factor Documentation for AP-42 Section 9.12.2: Wines and Brandy Final Report. US Environmental Protection Agency (1995) www.epa.gov/ttnchie1/ap42/ch09/bgdocs/b9s12-2.pdf

43. I. Vázquez-Rowe, P. Villanueva-Rey, D. Iribarren, M. Moreira, G. Feijoo. Joint life cycle assessment and data envelopment analysis of grape production for vinification in the Rías Baixas appellation (NW Spain). J. Clean. Prod., 27 (2012), pp. 92–102

44. I. Vázquez-Rowe, P. Villanueva-Rey, M. Moreira, G. Feijoo. Environmental analysis of Ribiero wine from a timeline perspective: harvest year matters when reporting environmental impacts. J. Environ. Manag., 98 (2012), pp. 73–83

45. P. Villanueva-Rey, I. Vázquez-Rowe, M. Moreira, G. Feijoo. Comparative life cycle assessment in the wine sector: biodynamic vs. conventional viticulture activities in NW Spain. J. Clean. Prod., 65 (2014), pp. 330–341

46. Wine Australia. Red Wine Overview (2012) www.wineaustralia.com/Australia/Default.aspx?tabid=806

47. Wine Australia. Grape Varieties (2013) www.wineaustralia.net.au/en/grape-varieties.aspx

48. Winebiz. Australasia's Wine Industry Portal by Wine Titles (2013) http://www.winebiz.com.au/statistics/wineries_branded.asp

49. WRAP. The Life Cycle Emissions of Wine Imported to the UK. Waste and Resources Action Programme, London, UK (2007) www.wfa.org.au/entwine_website/files/resources/The_Life_Cycle_Emissions_of_Wine_Imported_to_the_UK_Final_Report.pdf

50. WRAP. Bulk Shipping of Wine and its Implications for Product Quality. Waste and Resources Action Programme, London, UK (2008) www.wrap.org.uk/downloads/Bulk_shipping_wine_quality_May_08.d915ba2a.5386.pdf

PART II

FACTORS THAT IMPACT THE QUEST FOR SUSTAINABLE ENOLOGY

CHAPTER 3

MULTISTARTER FROM ORGANIC VITICULTURE FOR RED WINE MONTEPULCIANO D'ABRUZZO PRODUCTION

GIOVANNA SUZZI, MARIA SCHIRONE, MANUEL SERGI, ROSA MARIA MARIANELLA, GIUSEPPE FASOLI, IRENE AGUZZI, AND ROSANNA TOFALO

3.1 INTRODUCTION

Over the last decade, the phenomenon of organic products has taken hold of large segments of consumers. At its most basic level, organic wine is made from grapes that have been grown with as little human impact as possible. Organic wine is a wine obtained from organically growing grapes without the help of or need for synthetic fertilizers, synthetic plant treatments, or herbicides (Trioli and Hofmann, 2009). Accurate studies have been carried out on soil and vineyard management in organic wine making whereas, as far as we know, no or few data are available on microbial populations of grape berries from organic vineyard as well as of those from organic wines. In a previous paper (Tofalo et al., 2011) the yeast

Suzzi G, Schirone M, Sergi M, Marianella RM, Fasoli G, Aguzzi I and Tofalo R (2012) Multistarter from organic viticulture for red wine Montepulciano d'Abruzzo production. Frontiers in Microbiology *3:135. doi: 10.3389/fmicb.2012.00135. Reprinted with permission from the authors.*

populations present on grape berries and must from organic vineyards of red Montepulciano d'Abruzzo and white Trebbiano cultivars were studied. In particular non-*Saccharomyces* (NS) wine yeasts were identified at species level. Moreover the strains were typed and characterized for some oenological parameters. In recent years, a lot of studies evaluated the NS species present in wine ecosystem, and demonstrated the impact of grape conditions on NS populations (Fernández et al., 2000; Raspor et al., 2006; González et al., 2007). The role of NS yeasts in wine production has been debated extensively and several researchers have shown that NS yeasts survive during fermentation and could reach cell concentrations similar to those reached by *Saccharomyces cerevisiae* 10^6–10^8 cells/ml (Fleet et al., 1984; Gafner and Schultz, 1996). In fact, as suggested by several authors (Zironi et al., 1993; Gil et al., 1996; Lema et al., 1996; Toro and Vazquez, 2002; Ciani et al., 2006; Viana et al., 2008), there is growing evidence that NS yeasts play an important role in wine quality. Fleet (2008) discussed the possibilities of using yeasts other than those from the genus *Saccharomyces* for future wine fermentations and the commercial viability of mixed cultures, because NS species have great potential to introduce appealing characteristics to wine that may improve its organoleptic quality. Consequently, the impact of NS yeasts on wine fermentation cannot be ignored. The major NS yeasts present during organic must fermentation of Trebbiano and Montepulciano cultivars were *Hanseniaspora uvarum*, *Metschnikowia fructicola* and *Candida zemplinina*, representing 43, 31 and 11%, respectively, of the total NS population isolated. Although the population size of these species was reduced throughout the wine fermentations, their growth was not completely suppressed and NS yeasts were still present at the end of the fermentation process (Tofalo et al., 2011). These yeasts from organic wine shared many characteristics which suggest that the strong selection pressure exerted by farming system and vine variety could have generated variability at different levels. Knowledge about the biodiversity of native yeasts is essential for the preservation and exploitation of the oenological potential of wine grape growing regions. The use of a selected multistarter (controlled mixed cultures) was proposed several years ago. In the middle of the last century, to reduce the acetic acid content of wine, Cantarelli (1955), Castelli (1969) encouraged the sequential use of *Torulaspora delbrueckii* (formerly known as

Saccharomyces rosei) and *S. cerevisiae*. Later on, mixed cultures were proposed also for other objectives such as the biological deacidification of must or increase the glycerol content but one of the most investigated uses of mixed cultures relates to confer greater complexity to a wine, enhancing its organoleptic profile (Ciani et al., 2010). Several studies on mixed fermentations containing *S. cerevisiae* and NS wine yeasts have been carried out to evaluate the possibility of using controlled multistarter cultures to improve wine quality (Mora et al., 1990; Zironi et al., 1993; Toro and Vazquez, 2002; Ciani et al., 2006; Andorrà et al., 2010). For this purpose different NS species were used to study mixed fermentation such as *Hanseniaspora guilliermondii, H. uvarum, Candida pulcherrima* (*Metschnikowia*), *Pichia kluyveri, Pichia fermentans, Candida cantarellii, T. delbrueckii, Kluyveromyces thermotolerans, Candida stellata* (recently reclassified as *Starmerella bombicola*). The yeasts present on grape berries and must from organic vineyards could have a unique composition and these indigenous yeasts impart distinct regional and desired characteristics to wines. In this context, autochthonous NS strains could be selected to conferment organic musts alongside *S. cerevisiae*.

The aim of this research is to evaluate the fermentation performance and the interactions of mixed and sequential cultures of *H. uvarum* and *C. zemplinina* and a strain of *S. cerevisiae* isolated from organic must. The results of fermentation kinetics, secondary compound formation and sensorial analysis could be useful to formulate mixed starter cultures.

3.2 MATERIALS AND METHODS

3.2.1 YEAST STRAINS

Non-*Saccharomyces* strains (*C. zemplinina* STS12 and *H. uvarum* STS45) have been isolated in a previous study from spontaneous fermentation of organic Montepulciano d'Abruzzo and Trebbiano grapes (Tofalo et al., 2011). All of the strains have been previously analyzed by sequencing the approximately 600 bp D1/D2 region of the large (26S) ribosomal subunit using primers NL1 and NL4. These natural wine strains belong to the culture collection of the Food Science Department (University of Teramo,

Italy). An autochthonous starter culture *S. cerevisiae* (STS1) was also used as starter culture (Tofalo et al., 2011). The strains were maintained at −80°C in glycerol 20% (v/v) and, in parallel, on agar slants under paraffin oil at 4°C.

3.2.2 MICROVINIFICATIONS

To evaluate oenological performances, the strains were tested in micro-vinification trials using pasteurized must without grape skin from Montepulciano d'Abruzzo cultivar (280 g/l fermentable sugars, 7.4 g/l titratable acidity (TTA) and a pH 3.2). Fermentations were conducted using several combinations of the strain *S. cerevisiae* STS1 with strains *C. zemplinina* STS12 and *H. uvarum* STS45. The musts were inoculated with 10^6 cells/ml as follows: *S. cerevisiae* (S), *C. zemplinina* (C), *H. uvarum* (H), *C. zemplinina/S. cerevisiae* (CS), *H. uvarum/S. cerevisiae* (HS), *C. zemplinina/H. uvarum/S. cerevisiae* (CHS), moreover the must was inoculated with all the three strains as follows *C. zemplinina/H. uvarum/S. cerevisiae* (RCHS) ratios of 25:25:50. A sequential fermentation (QS) was inoculated with 10^6 cells/ml *C. zemplinina/H. uvarum* and after 48 h *S. cerevisiae* was added for each strain.

The must samples (95 ml) were inoculated with 5 ml of a pre-culture grown for 48 h in the same must, as described by Tofalo et al. (2011). Fermentations were carried out in duplicate for each strain at a controlled temperature of 25°C. The kinetic fermentations were monitored daily by gravimetric determinations, evaluating the loss of weight due to the production of CO_2. When the CO_2 evolution stopped (i.e., at constant weight), the samples were refrigerated for 2 days at 4°C, racked, and stored at −20°C until analysis. Non-inoculated must was used as negative control.

3.2.3 DETERMINATION OF MICROBIAL GROWTH AND DIFFERENTIAL ENUMERATION

From each flask, samples were taken along the fermentation process to evaluate viable cell counts of the inoculated species. One hundred microliters aliquots of serial dilutions of each sample were plated on Lysine Agar (LA medium; Oxoid Unipath, Hampshire, UK) and Wallerstein Laboratory nutrient agar medium (WLN medium; Oxoid; Pallmann et al., 2001) to estimate the NS yeast and the total yeast population, respectively. Moreover, a culture-independent approach was used. *S. cerevisiae, H. uvarum, C. zemplinina* specific quantitative PCR (qPCR) tests were carried out according to Hierro et al. (2007) and Zott et al. (2010), respectively. DNA was extracted using the DNA PowerSoil® Isolation Kit (Mobio Laboratories, Inc.) according to the manufacturer's instructions.

Real-time amplifications were carried out in a 25 µl reaction volume, using 1× 2XIQ SYBR Green PCR Supermix (Bio-Rad, Hercules, CA, USA), 0.2 µM of each primer and 5 µl of DNA suspension. All amplifications were carried out in optical-grade 96-well plates on a Cycle IQ system (Bio-Rad). The qPCR threshold cycle (C_t) was determined automatically by the instrument. Samples with known quantities of yeast cells were prepared to generate standard curves. Sterilized grape juice was inoculated with the yeast strains, plated on WLN media for viable counts. The counted samples were immediately extracted (triplicate), as described above. The DNA obtained was used to prepare serial dilutions, from 10^8 to 10 cell/ml. The correlation coefficient between C_t and count values was analyzed and interpreted using the appropriate Microsoft Excel function. Each C_t was the average of four measures obtained by amplifying four DNA extracts from the same artificially inoculated sample. In all PCR runs, negative controls (sterilized water), positive controls, and samples were run in triplicate. Sensitivity of qPCR assays was evaluated with reference to other reports (Hierro et al., 2006).

3.2.4 PHYSICO-CHEMICAL DETERMINATIONS

The main products (ethanol, glycerol, pH, TTA, sugar content, free sulfur dioxide (SO_2) and dry extract of wine), and must under fermentation were determined on samples taken at the end of fermentation following the official International Organization of Vine and Wine (2011) methods of analysis. Organic acid, glucose, and fructose concentrations were determined according to Tofalo et al. (2011) and Lopez-Tamames et al. (1996), respectively. Biogenic amines production was determined according to Tofalo et al. (2007). All analyses were performed in triplicate.

3.2.5 SOLID PHASE MICROEXTRACTION–GAS CHROMATOGRAPHY ANALYSIS OF VOLATILE COMPOUNDS

Five milliliters of wine samples were placed in 10 ml glass vials with 1 g NaCl and 10 μl of 4-methyl-2-pentanol (final concentration 4 mg/l) were added as internal standard. Both equilibration and adsorption phases were carried out by stirring for 30 min at 40°C. A carboxen–polydimethylsiloxane-coated fiber (85 μm) was used (Sigma-Aldrich, St. Louis, MO, USA). Under the extraction conditions described above, the recovery of the volatile compounds was between 88.9 and 103.5%. For quantitative determination, a CP 380 capillary gas chromatograph equipped with a 8200 autosampler SPME III (Varian, Italy) was used. The fused silica capillary column was a CP-Wax 52 CB (50 m × 0.32 mm) by Crompack (Netherlands), coated with polyethylene glycol (film thickness 1.2 μm), as stationary phase. The injector and FID temperature was 250°C. After extraction, the fiber was placed in the injector of the GC for 15 min. The temperature program was the following: initial temperature (50°C) held for 2 min; first ramp, 1°C min to 65°C (0 min hold); second ramp, 10°C min to 150°C (10 min hold); third ramp 10°C min to 200°C (1 min hold). The carrier gas (N_2) flow rate was 2.5 ml/min. The aroma compounds were identified by comparing the retention time of standards and their identification was confirmed by using GC–MS. GC–MS analysis was performed using a GC–mass spectrometer Finnigan Trace DSQ Quadrupole (Thermo Finnigan, San Jose, CA, USA). Mass spectrometer conditions were: Ion Source: electron ionization (EI), Ion Polarity: POS, Ion Source Temperature: 250°C, MS transfer line: 220°C, Turbomolecular Pump: 70 l/s, Acquisition: full Scan, Mass range: 30–400 m/z, Carrier gas: He. The data were processed using Xcalibur Data System Software 1.4.1 SP3 (Thermo Finnigan, San Jose, CA, USA). The quantitative analysis of wine aroma compounds was carried out on the basis of the relative peak area (Q_i) calculated from head space SPME (HS/SPME) gas chromatograms after addition of know amounts of analyte standards, as well as the internal standard according to De la Calle-Garcia et al. (1998). The chemical analyses were carried out in the same period of the sensory analysis. Each determination was carried out in duplicate. The data presented are the means of three determinations. All reagents were purchased from Sigma-Aldrich (St. Louis, MO, USA), with a purity greater than 99%.

3.2.6 STATISTICAL ANALYSIS

The mean and SD were calculated for each experimental parameter. Principal component analysis (PCA) was performed using statistical software STATISTICA for Windows (STAT. version 8.0, StatSoft Inc., Tulsa, OK, USA).

3.3 RESULTS

3.3.1 COURSE OF FERMENTATION AND DEVELOPMENT OF YEASTS DURING FERMENTATIONS

The course of fermentation rates with pure, mixed, and sequential cultures is given in Figure 1. The pure culture of *H. uvarum* was unable to finish fermentation according to the poor fermentative capacity of this species, but also the pure culture of *C. zemplinina* did not end the fermentation. The yeast interactions had a clear impact on the fermentation kinetics and the presence of *S. cerevisiae* gave faster fermentations.

In fact, fermentation kinetics of mixed cultures, HS, CHS, RCHS were comparable to those of *S. cerevisiae* pure culture. Only the trial CS showed a great improvement of fermentation kinetic during the first 15 days of fermentation highlighting the good association between *C. zemplinina* and *S. cerevisiae*. The RCHS trial, characterized by the inoculum of *C. zemplinina* and *H. uvarum* together with *S. cerevisiae*, showed a worst fermentation kinetic than CS trial probably for the lower proportion of cells of the two NS with respect to *S. cerevisiae*. The sequential trial QS presented a dramatic decrease of fermentation rate probably because *S. cerevisiae* was inoculated 48 h after the inoculum of *C. zemplinina* and *H. uvarum*.

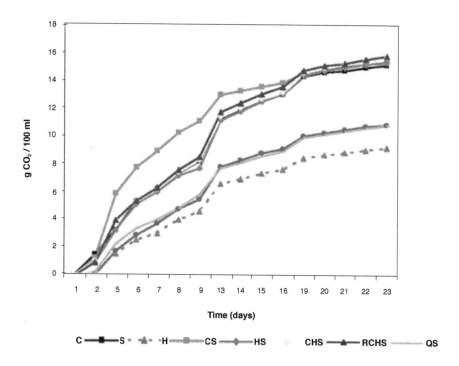

FIGURE 1: Fermentation kinetics of pure, mixed, and sequential starter cultures in Montepulciano d'Abruzzo musts. C, *C. zemplinina*; S, *S. cerevisiae*; H, *H. uvarum*; CS, *C. zemplinina/S. cerevisiae*; HS, *H. uvarum/S. cerevisiae*; CHS, *C. zemplinina/H. uvarum/S. cerevisiae*; RCHS, *C. zemplinina/H. uvarum/S. cerevisiae*; QS, sequential fermentation.

TABLE 1: Yeast counts (log CFU/ml) and quantification by qPCR (in brackets) in pure, mixed, and sequential fermentation of organic Montepulciano d'Abruzzo musts.

Trial	Time (days	Strains		
		S. cerevisiae	*C. zemplinina*	*H. uvarum*
C	0		6.79 ± 0.05*	
	2		7.09 ± 0.04	
	14		7.36 ± 0.04	
	24		5.64 ± 0.06	
S	0	6.88 ± 0.03		
	2	7.21 ± 0.03		
	14	7.27 ± 0.01		
	24	7.15 ± 0.04		
H	0			6.68 ± 0.2
	2			6.87 ± 0.03
	14			5.32 ± 0.03
	24			nd
CS	0	6.82 ± 0.04	6.85 ± 0.02	
	2	6.50 ± 0.02	6.08 ± 0.11	
	14	6.74 ± 0.07	nd	
	24	7.11 ± 0.04	nd	
HS	0	6.89 ±0.04		6.68 ± 0.02
	2	6.67 ± 0.02		5.32 ± 0.03
	14	7.08 ± 0.02		nd
	24	6.78 ± 0.0014		nd
CHS	0	6.88 ± 0.03 (6.98 ± 0.03	6.81 ± 0.02 (6.34 ± 0.03)	6.67 ± 0.01 (6.45 ± 0.04)
	2	6.87 ± 0.03 (7.24 ± 0.01)	6.69 ± 0.03 (6.45 ± 0.09)	6.29 ± 0.02 (6.39 ± 0.01)
	14	7.09 ± 0.04 (7.48 ±0.04)	nd (4.03 ± 0.03)	nd (5.28 ± 0.01)
	24	6.72 ±0.03 (7.48 ± 0.01)	nd (3.76 ± 0.01)	nd (4.83 ± 0.02)
RCHS	0	6.88 ± 0.03 (6.98 ± 0.04)	3.84 ± 0.01 (3.36 ± 0.05)	3.68 ± 0.01 (3.50 ± 0.11)
	2	7.08 ± 0.02 (7.24 ± 0.01)	5.06 ± 0.08 (5.01 ± 0.01)	nd (3.58 ± 0.03)
	14	7.44 ± 0.02 (7.48 ± 0.04)	5.04 ± 0.04 (5.09 ± 0.04)	nd (5.60 ± 0.07)
	24	7.46 ± 0.04 (7.48 ± 0.01)	nd (4.91 ± 0.02)	nd (5.40 ± 0.09)

TABLE 1: *Cont.*

Trial	Time (days	Strains		
		S. cerevisiae	*C. zemplinina*	*H. uvarum*
QS	0	–	6.85 ± 0.02 (6.45 ± 0.02)	6.68 ± 0.02 (6.34 ± 0.03)
	2	6.9 ± 0.01 (6.98 ± 0.01)	7.16 ± 0.03 (7.24 ± 0.01)	7.19 ± 0.01 (7.35 ± 0.02)
	14	6.14 ± 0.04 (6.15 ± 0.02)	6.89 ± 0.01 (6.90 ± 0.04)	nd (4.43 ± 0.01)
	24	5.92 ± 0.04 (7.16 ± 0.01)	6.08 ± 0.02 (6.29 ± 0.01)	nd (4.05 ± 0.1)

*C, C. zemplinina; S, S. cerevisiae; H, H. uvarum; CS, C. zemplinina/S. cerevisiae; HS, H. uvarum/S. cerevisiae; CHS, C. zemplinina/H. uvarum/S. cerevisiae; RCHS, C. zemplinina/H. uvarum/ S. cerevisiae; QS, sequential fermentation. *Data are expressed as average ± SD; nd, not inoculated.*

The viable counts of yeast populations of the eight trials are reported in Table 1. The yeast populations of pure cultures of *S. cerevisiae* and *C. zemplinina* were similar, reaching maximum population around 10^7 CFU/ml after 48 h. In the case of *H. uvarum* pure culture the maximum population (around 106 CFU/ml) was reached after 48 h, but starting from the 14th day it started to decrease at 10^5 CFU/ml and was not more countable on the 24th day. Regarding the mixed fermentations HS and CS, NS yeasts were not found at the 14th day, whereas *S. cerevisiae* delayed the reaching of maximum population. As regards the viable cells of the three species during mixed fermentation CHS and RCHS, differences were observed only in the latest one. In fact *H. uvarum* disappeared after 48 h, whereas *C. zemplinina* increased up to 10^5 CFU/ml and *S. cerevisiae* reached maximum population of about 10^7 CFU/ml on the 14th day. In general, *S. cerevisiae* quickly reached its maximum population and kept stable during the vinifications, with the exception of sequential fermentation QS during which *S. cerevisiae* was not able to grow after the 48 h growth of NS wine yeast. In fact, the number of viable cells of *S. cerevisiae* decreased up to 10^5 CFU/ml at 24th day. *H. uvarum* and *C. zemplinina* reached the maximum population of 10^7 CFU/ml instead after 48 h, but during the progress of fermentation the first one was not more countable whereas *C. zemplinina* decreased up to 10^6 CFU/ml.

3.3.2 QUANTIFICATION OF MULTISTARTER POPULATION DURING FERMENTATION BY qPCR

As culture-dependent techniques can underestimate the size and the diversity of a given population because they do not account for non-cultivable populations, the data on *S. cerevisiae* and NS populations during mixed fermentation obtained by plating were compared with those obtained by qPCR, as reported in Section "Materials and Methods." In particular the trials CHS, RCHS, and QS were analyzed by real-time PCR and the obtained results are reported in Table 1. *S. cerevisiae* resulted to be always present at high level (10^6–10^7 cells/ml), but also *H. uvarum* and *C. zemplinina* were found at the end of fermentation, generally at about 10^4 cells/ml. In particular *H. uvarum* at the end of fermentation of trail CHS showed a Ct value of 25, corresponding to 10^4 cells/ml (data not shown).

3.3.3 MAIN FERMENTATION PRODUCTS AND VOLATILE COMPOUNDS

Table 2 reports some oenological parameters of the pure culture, mixed, and sequential fermentations. The pure cultures of *H. uvarum* and *C. zemplinina* did not finish the fermentations, leaving in the medium glucose and fructose, and only glucose, respectively. Also trial QS had 50 g/l of residual glucose, as expected by the analysis of fermentation kinetics. The highest ethanol concentration was determined in pure culture of *S. cerevisiae*. All the multistarter fermentations reached a lower ethanol concentration ranging from 8.34 to 9.38%. On the contrary the production of glycerol was greater by NS yeasts and in mixed cultures. Despite the high production of acetic acid in pure culture, *H. uvarum* did not increase volatile acidity in multistarter fermentations. No significant differences were found in the other compounds. Some discrepancies were found among ethanol, residual sugars, and other secondary metabolic compounds. What is stricking in our experiments was that all the strains showed an extremely poor ethanol yield from sugar consumed, which cannot be explained by the overproduction of any other metabolic products investigated in this study. Similar results have been previously obtained by Magyar and Toth (2011) and Tofalo et al. (2012).

Biogenic amines were detected at low levels in all eight microvinifications (Figure 2). In particular, the trial CS has the highest value of biogenic amines and QS has the lowest value. The pure cultures of *H. uvarum*, *C. zemplinina*, and *S. cerevisiae* did not produce cadaverine that was formed in all multistarter fermentations with the exception of QS. Tyramine, histamine, 2-phenylethylamine, spermidine, and methylamine were not detectable in any the samples analyzed.

Table 3 shows some volatile compounds that well discriminated the aromatic profiles of the three wine yeast species. Pure cultures of *C. zemplinina* and *S. cerevisiae* produced low quantities of ethyl acetate and acetoin and high amounts of isoamyl alcohols and β-phenylethanol, having *C. zemplinina* the highest productions with respect to *S. cerevisiae*. In this study *H. uvarum* pure culture produced high amounts of ethyl acetate and acetoin and very low amounts of isoamyl alcohols and β-phenylethanol. The fermentations of mixed cultures CS, HS, and RCHS presented low levels of ethyl acetate and decreased the quantity of isoamylic alcohols and 2-phenylethanol with respect to *S. cerevisiae*. The fermentation of mixed culture CHS presented the highest formation of isoamyl alcohols and ethyl acetate. The sequential fermentation QS showed a completely different situation among the multistarter fermentations, with the lowest content of ethyl acetate and isoamyl alcohols and the highest one of acetoin, 2-phenylethanol, and isobutyl alcohol.

Principal component analysis has been used to obtain biochemical fingerprints of wines and to elucidate differences in the different components of wine fermented by different strains of *Saccharomyces* or by multistarter cultures (Nurgel et al., 2002; Howell et al., 2006).

Some compounds produced by the yeast pure and multistarter cultures were analyzed using PCA (Figures 3A,B). Firstly, the correlation matrix was computed in order to discriminate the variables, thus selecting 10 parameters (acetoin, 2,3 butanediol, ethanol, ethyl acetate, titratable and volatile acidity, glycerol, reducing sugars, phenylethyl alcohol and isoamyl alcohols). The PCA explained 70.6% of the total variance. PC 1 accounted for 53.55% of the variance and the negative segment of loading plot for this dimension (Figure 3B) was closely related to the levels of volatile acidity, whereas its positive counterpart was mainly related to ethanol. PC2 explained 17.05% of the variance; this dimension was mainly related positively with titratable acidity. Then, in score plot (Figure 3A) it is possible to distinguish three different groups of wines produced by strains. The fermentation carried out by *H. uvarum* was well differentiated from the others for the highest concentration of ethyl acetate, glycerol, acetoin, and volatile acidity; whereas C, CS, and CHS were very similar except for the concentration of isoamyl alcohols, 2,3 butanediol and reducing sugars (in C and CS, the values were higher than in CHS).

3.4 DISCUSSION

One of the most studied technologically advances in wine making is the inoculation of grape juice with mixed cultures of *S. cerevisiae* and NS yeasts. In this study the fermentation kinetics and metabolic compounds produced by multistarters during fermentation of organic musts were compared. Generally the fermentation kinetics of pure cultures of *S. cerevisiae*, *H. uvarum*, and *C. zemplinina* were in agreement with those expected and reported in other studies (Egli et al., 1998; Toro and Vazquez, 2002; Zohre and Erten, 2002; Mendoza et al., 2007; Fleet, 2008; Ciani et al., 2010). The pure culture of *H. uvarum* was unable to finish fermentation according to the poor fermentative capacity of this species, but also the pure culture of *C. zemplinina* did not end the fermentation. These two strains were selected on the basis of their fermentative capacity in a must containing 180 g/l sugars (Tofalo et al., 2011) and probably the metabolic fermentative products from the high sugar Montepulciano must (280 g/l) affected their performances. In particular, *C. zemplinina* has been reported to complete fermentation of Macabeo must containing 180 g/l sugars (Andorrà et al., 2010) even if with a slight delay compared to the *S. cerevisiae* fermentation (Sipiczki et al., 2005; Tofalo et al., 2009, 2012). The yeast interactions had a clear impact on the fermentation kinetics and the presence of *S. cerevisiae* gave faster fermentations. However *C. zemplinina* and *S. cere-*

visiae association (trial CS) showed a great improvement of kinetic during the first 15 days of fermentation. *C. zemplinina* is an osmotolerant and fructophilic yeast, generally producing low amounts of acetic acid and relevant quantities of glycerol from sugar fermentation (Sipiczki et al., 2005; Magyar and Toth, 2011; Tofalo et al., 2011), it could be suggested that it will be able to consume sugars at the very beginning of the fermentation, alleviating the *S. cerevisiae* from the osmotic stress, thereby improving also the fermentation kinetics (Rantsiou et al., 2012). As previously demonstrated for *S. cerevisiae* species, different strains of *C. zemplinina* can have a specific effect, as reported by Tofalo et al. (2012). Moreover this positive interaction on fermentation kinetics between *S. cerevisiae* and *C. zemplinina* can have an important application on must with high sugar contents or special wines such as icewine, "passito," botrytized wines (Rantsiou et al., 2012). On the contrary the inoculum of *C. zemplinina* and *H. uvarum* together with *S. cerevisiae* (RCHS) did not produce the same results of CS trial, probably for the lower proportion of cells of the two NS with respect to *S. cerevisiae*.

The inoculum of *S. cerevisiae* after 48 h since that of *C. zemplinina* and *H. uvarum* (trial QS) produced a stuck fermentation Similar results have been reported in other studies, even if the mechanisms of this performance reduction of *S. cerevisiae* have not yet been explained. When *S. cerevisiae* and C. cantarelli interact during fermentation, the maximum *S. cerevisiae* population decreases (Toro and Vazquez, 2002). This fact may be due to amino acid and vitamin consumption during the first days of fermentation that can disable the subsequent growth and fermentative capacity of *S. cerevisiae* (Fleet, 2003). On the other hand these data were confirmed by the plating counts. *S. cerevisiae* dominated all the multistarter fermentations a part trial QS in which *C. zemplinina* had its best performance. The higher ability of *S. cerevisiae* to withstand the stress conditions such as increasing ethanol, decreasing pH, nutrition depletion is generally considered to drive the wine yeast population dynamics (Pretorius, 2000). These selective pressures are currently questioned, whereas cell–cell interactions are being put forward as significant in affecting yeast succession (Ciani and Pepe, 2002; Fleet, 2003; Nissen et al., 2003). The early death of *C. zemplinina* and *H. uvarum* during mixed fermentation, even if with quantitative differences, appeared to be due to the antagonistic effect of *S. cerevisiae* as reported also by Andorrà et al. (2010) in the same species. The NS, in mixed fermentations, decreased as fermentation proceeded. On the contrary other authors (Ciani et al., 2006; Mendoza et al., 2007) reported that the presence of both *Saccharomyces* and NS yeasts promotes an increase in the persistence of NS yeasts during fermentation process. However it is clear that *S. cerevisiae* has a relevant antagonist effect upon *C. zemplinina* and *H. uvarum*, but also that the same NS can affect *S. cerevisiae*, depending on the sequence of growth, the number of viable cells and strain specificity. Few studies have been carried out to elucidate the mechanisms of these antagonistic phenomena (Nissen et al., 2003; Arneborg et al., 2005; Pérez-Nevado et al., 2006). During alcoholic fermentation yeasts can produce compounds that can have inhibitory effects against other yeast species or strains, such as short to medium-chain fatty acids (Ludovico et al., 2001; Fleet, 2003), killer toxins (Schmitt and Breinig, 2002), growth and nutritional conditions (Fleet, 2003). However in the above-mentioned studies, cell–cell contact-mediated mechanism appear to effect antagonism among yeasts. To confirm the data obtained by plating counts qPCR was used as culture-independent method. In fact sub lethally injured and/or viable but non-culturable (VBNC) cells, may fail to grow on plates and are common in wine (Millet and Lonvaud-Funel, 2000; Andorrà et al., 2011). The qPCR method used was previously developed to monitor the yeast evolution during alcoholic fermentations (Hierro et al., 2007; Andorrà et al., 2011). However, qPCR targeted at DNA quantifies also dead yeasts because of the DNA's stability (Hierro et al., 2006). Our data confirmed those of Andorrà et al. (2011) who found high populations of *H. uvarum* (up to 108 cells/ml) throughout the mixed fermentations by qPCR methods. Of course these findings do not resolve the question if these populations correspond to VBNC, injured and/or dead cells or how many of these cells are metabolically active and how these in turn can influence the final wines (Andorrà et al., 2011). This aspect needs further researches and advances. When some yeasts develop together under fermentation conditions, they do not passively coexist, but rather they interact and can produce different levels of fermentation products, which can affect the chemical and aromatic composition of wines (Howell et al., 2006; Anfang et al., 2009). Grape must fermentations performed by pure, mixed or sequential cultures of NS yeasts with *S. cerevisiae* can produce wines with significant differences in the chemical composition (Herraiz et al., 1990; Ciani and Picciotti, 1995; Gil et al., 1996; Lambrechts and Pretorius, 2000; Rojas et al., 2003; Romano et al., 2003; Moreira et al., 2008). A relevant consequence of the NS yeasts was the increased presence of glycerol in all the multistarter fermentations, due the intrinsic characteristic of *C. zemplinina* and *H. uvarum*. In similar way acetic acid production is considered as a common pattern in apiculate yeasts that for this reason have been considered for long time as spoilage yeasts (Romano et al., 2003). Despite the high production of acetic acid in pure culture, *H. uvarum* did not increase volatile acidity in multistarter fermentations. These results are in complete agreement with those reported by other authors (Ciani et al., 2006; Mendoza et al., 2007; Andorrà et al., 2010). As regards biogenic amines, few studies have been conducted on their formation by yeasts, comparing different yeast species and quantifying only histamine (Torrea and Ancín, 2002). Their presence in wine ranged from a few milligrams per liter to about 50 mg/l depending on the wine. This study confirmed the low amino-decarboxylase of wine yeasts, even if some differences were determined in the eight trials. Even if a number of authors suggest that yeasts do not appear to be responsible for the production of most amines found in industrial commercial red wines (Marcobal et al., 2006; Smit et al., 2008), the yeast contribution to biogenic amine production could therefore be indirect: yeasts can alter the composition of grape musts by using some amino acids and secreting others during alcoholic fermentation and autolysis, thereby changing the concentration of precursor amino acids in the wine that can be used by other microorganisms in subsequent fermentation steps (Soufleros et al., 1998). According to the intrinsic characteristic of the single species, the production of main aromatic compounds was well differentiated in the trials. Low quantities of ethyl acetate and acetoin and high amounts of isoamyl alcohols and β-phenylethanol were produced by pure cultures of *C. zemplinina* and *S. cerevisiae*, as similarly reported by Andorrà et al. (2010). In this study *H. uvarum* pure culture produced high amounts of ethyl acetate and acetoin and very low amounts of isoamyl alcohols and β-phenylethanol. The production of large quantities of ethyl acetate and acetic acid by *H. uvarum* has always been considered a negative characteristic (Ciani and Picciotti,

1995), whereas there is controversy concerning the production of higher alcohols. *H. uvarum* and K. apiculata were found to be the main producers of higher alcohols by Gil et al. (1996). However, some authors (Herraiz et al., 1990; Rojas et al., 2003; Romano et al., 2003) reported that apiculate yeasts were low producers of higher alcohols and can promote the esterification of various alcohols such as ethanol, geraniol, isoamyl alcohols, and 2-phenylethanol. Probably high production of higher alcohols could be a strain character dependent also in some NS wine yeasts (Capece et al., 2005). The secondary volatile compounds produced by mixed cultures were a combination of the different strains. PCA well highlighted these differences, indicating that the final wine characteristics can be modulated by different combinations or species and sequence of inoculum (Nurgel et al., 2002; Howell et al., 2006).

Candida zemplinina showed interesting features for mixed fermentation with *S. cerevisiae*, in particularly increasing the fermentation kinetic in high gravity Montepulciano must, with low ethyl acetate and acetic acid production. The combined use of three starter cultures (CSH and RCSH) could allow the improvement of the organoleptic characteristics of a wine. Further studies are needed to clarify the interaction among the different starters and to optimize the fermentation and the modalities of inoculation. Effectively the use of NS wine yeasts together with *Saccharomyces* strains in mixed fermentations might be recommended as a tool to obtain the advantages of spontaneous fermentation of organic wines such as those obtained with Montepulciano, while avoiding the risks of stuck fermentation (Rojas et al., 2003; Romano et al., 2003; Jolly et al., 2006; Ciani et al., 2010). Because of the NS yeast strains biodiversity about their production level of enzymatic activities (Manzanares et al., 1999, 2000; Mendes-Ferreira et al., 2001; Strauss et al., 2001) and fermentation metabolites (Romano et al., 1992, 2003; Capece et al., 2005) of enological importance, suitable strains should be selected in order to be able to design mixed starter able to provide beneficial contributions also for wine obtained by organic viticulture. In particular *C. zemplinina* showed a good interaction with *S. cerevisiae* by increasing the fermentation kinetic in high gravity Montepulciano must, with low ethyl acetate and acetic acid production.

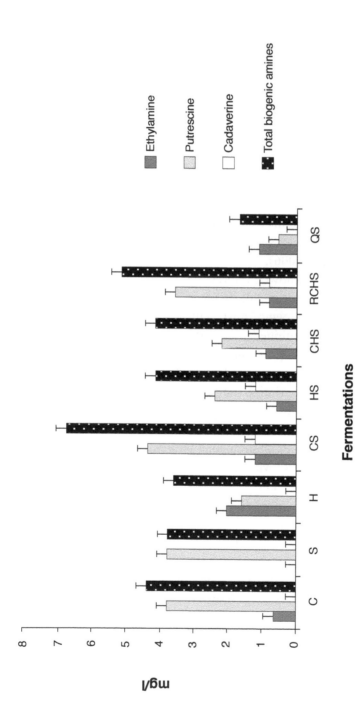

FIGURE 2: Biogenic amines content in wines obtained with pure, mixed, and sequential starter cultures. C, *C. zemplinina*; S, *S. cerevisiae*; H, *H. uvarum*; CS, *C. zemplinina/S. cerevisiae*; HS, *H. uvarum/S. cerevisiae*; CHS, *C. zemplinina/H. uvarum/S. cerevisiae*; RCHS, *C. zemplinina/H. uvarum/S. cerevisiae*; QS, sequential fermentation.

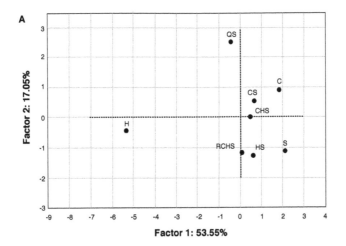

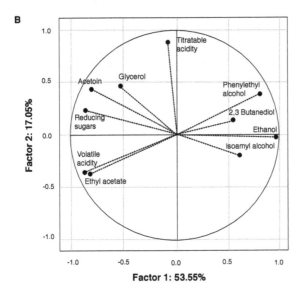

FIGURE 3: Score plot (A) and loading plot (B) of the first and second principal components (PC) after PC analysis by the yeast pure and multistarter cultures. C, *C. zemplinina*; S, *S. cerevisiae*; H, *H. uvarum*; CS, *C. zemplinina/S. cerevisiae*; HS, *H. uvarum/S. cerevisiae*; CHS, *C. zemplinina/H. uvarum/S. cerevisiae*; RCHS, *C. zemplinina/H. uvarum/S. cerevisiae*; QS, sequential fermentation.

REFERENCES

1. Andorrà, I., Berradre, M., Rozès, N., Mas, A., Guillamón, J. M., and Esteve-Zarzoso, B. (2010). Effect of pure and mixed cultures of the main wine yeast species on grape must fermentations. Eur. Food Res. Technol. 231, 215–224.

2. Andorrà, I., Monteiro, M., Esteve-Zarzoso, B., Albergaria, H., and Mas, A. (2011). Analysis and direct quantification of Saccharomyces cerevisiae and Hanseniaspora guilliermondii populations during alcoholic fermentation by fluorescence in situ hybridization, flow cytometry and quantitative PCR. Food Microbiol. 28, 1483–1491.

3. Anfang, N., Brajkovich, M., and Goddard, M. R. (2009). Co-fermentation with Pichia kluyveri increases varietal thiol concentrations in Sauvignon Blanc. Aust. J. Grape Wine Res. 15, 1–8.

4. Arneborg, N., Siegumfeldt, H., Andersen, G., Nissen, P., Daria, V., Rodrigo, P., and Gluckstad, J. (2005). Interactive optical trapping shows that confinement is a determinant of growth in a mixed yeast. FEMS Microbiol. Lett. 245, 155–159.

5. Cantarelli, C. (1955). Studio comparativo dei lieviti apiculati dei generi Kloeckera (Janke) ed Hanseniaspora (Zikes). Ann. Microbiol. 6, 85.

6. Capece, A., Fiore, C., Maraz, A., and Romano, P. (2005). Molecular and technological approaches to evaluate strain biodiversity in Hanseniaspora uvarum of wine origin. J. Appl. Microbiol. 98, 136–144.

7. Castelli, T. (1969). Vino al microscopio. Roma: Scialpi Editore.

8. Ciani, M., Beco, L., and Comitini, F. (2006). Fermentation behaviour and metabolic interactions of multistarter wine yeast fermentations. Int. J. Food Microbiol. 108, 239–245.

9. Ciani, M., Comitini, F., Mannazzu, I., and Domizio, P. (2010). Controlled mixed culture fermentation: a new perspective on the use of non-Saccharomyces yeasts in winemaking. FEMS Yeast Res. 10, 123–133.

10. Ciani, M., and Pepe, V. (2002). The influence of pre-fermentative practices on the dominance of inoculated yeast starter under industrial conditions. J. Sci. Food Agric. 82, 573–578.

11. Ciani, M., and Picciotti, G. (1995). The growth kinetics and fermentation behaviour of some non-Saccharomyces yeasts associated with wine-making. Biotechnol. Lett. 17, 1247–1250.

12. De la Calle-Garcia, D., Reichenbacher, M., Dancer, K., Hurlbeck, C., Bartzsch, C., and Feller, K. H. (1998). Analysis of wine bouquet components using headspace space solid-phase microextraction-capillary gas chromatography. J. High Resolut. Chromatogr. 21, 373–377.

13. Egli, C. M., Edinger, W. D., Mitrakul, C. M., and Henick-Kling, T. (1998). Dynamics of indigenous and inoculated yeast populations and their effect on the sensory character of riesling and Chardonnay wines. J. Appl. Microbiol. 85, 779–789.

14. Fernández, M. T., Ubeda, J. F., and Briones, A. I. (2000). Typing of non-Saccharomyces yeasts with enzymatic activities of interest in winemaking. Int. J. Food Microbiol. 59, 29–36.

15. Fleet, G. H. (2003). Yeast interactions and wine flavour. Int. J. Food Microbiol. 86, 11–22.

16. Fleet, G. H. (2008). Wine yeast for the future. FEMS Yeast Res. 8, 979–995.
17. Fleet, G. H., Lafon-Lafourcade, S., and Ribéreau-Gayon, P. (1984). Evolution of yeasts and lactic acid bacteria during fermentation and storage of Bordeaux Wines. Appl. Environ. Microbiol. 48, 1034–1038.
18. Gafner, J., and Schultz, M. (1996). Impact of glucose-fructose ratio on stuck fermentations: practical experiences to restart stuck fermentation. Vitic. Enol. Sci. 51, 214–218.
19. Gil, J., Mateo, J., Jiménez, M., Pastor, A., and Huerta, T. (1996). Aroma compounds in wine as influenced by apiculate yeasts. J. Food Sci. 61, 1247–1266.
20. González, S. S., Barrio, E., and Querol, A. (2007). Molecular identification and characterization of wine yeast isolated from Tenerife (Canary Island, Spain). J. Appl. Microbiol. 102, 1018–1025.
21. Herraiz, T., Reglero, G., Herraiz, M., Martín-Alvarez, P., and Cabezudo, M. D. (1990). The influence of the yeast and type of culture on the volatile composition of wines fermented without sulfur dioxide. Am. J. Enol. Vitic. 41, 313–318.
22. Hierro, N., Esteve-Zarzoso, B., González, A., Mas, A., and Guillamón, J. M. (2006). Real-time quantitative PCR (QPCR) and reverse transcription-QPCR for detection and enumeration of total yeasts in wine. Appl. Environ. Microbiol. 72, 7148–7155.
23. Hierro, N., Esteve-Zarzoso, B., Mas, A., and Guillamón, J. M. (2007). Monitoring of Saccharomyces and Hanseniaspora populations during alcoholic fermentation by real-time quantitative PCR. FEMS Yeast Res. 7, 1340–1349.
24. Howell, K. S., Cozzolino, D., Bartowsky, E. J., Fleet, G. H., and Henschke, P. A. (2006). Metabolic profiling as a tool for revealing Saccharomyces interactions during wine fermentation. FEMS Yeast Res. 6, 91–101.
25. International Organization of Vine and Wine. (2011). Compendium of International Methods of Wine and must Analysis.
26. Jolly, N. P., Augustyn, O. P. H., and Pretorius, I. S. (2006). The role and use of non-Saccharomyces yeasts in wine production. S. Afr. J. Enol. Vitic. 27, 15–39.
27. Lambrechts, M. G., and Pretorius, I. S. (2000). Yeast and its importance to wine aroma. A review. South Afr. J. Enol. Vitic. 21, 97–129.
28. Lema, C., Garcia-Jares, C., Orriols, I., and Angulo, L. (1996). Contribution of Saccharomyces and non-Saccharomyces populations to the production of some components of Albarino wine aroma. Am. J. Enol. Vitic. 47, 206–216.
29. Lopez-Tamames, E., Puig-Deu, M., Teixeira, E., and Buxaderas, S. (1996). Organic acids, sugars, and glycerol content in white winemaking products determined by HPLC: relationship to climate and varietal factors. Am. J. Enol. Vitic. 47, 193–198.
30. Ludovico, P., Sousa, M. J., Silva, M. T., Leao, C., and Corte-Real, M. (2001). Saccharomyces cerevisiae commits to a programmed cell death process in response to acetic acid. Microbiology 147, 2409–2415.
31. Magyar, I., and Toth, T. (2011). Comparative evaluation of some oenological properties in wine strains of Candida stellata, Candida zemplinina, Saccharomyces uvarum and Saccharomyces cerevisiae. Food Microbiol. 28, 94–100.
32. Manzanares, P., Ramon, D., and Querol, A. (1999). Screening of non-Saccharomyces wine yeasts for the production of β-D-xylosidase activity. Int. J. Food Microbiol. 46, 105–112.

33. Manzanares, P., Rojas, V., Genovés, S., and Vallés, S. (2000). A preliminary search for anthocyanin-β-D-glucosidase activity in non-*Saccharomyces* wine yeasts. Int. J. Food Sci. Technol. 35, 95–103.

34. Marcobal, Á., Martín-Álvarez, P. J., Polo, C., Muñoz, R., and Moreno-Arribas, M. V. (2006). Formation of biogenic amines throughout the industrial manufacture of red wine. J. Food Prot. 69, 397–404.

35. Mendes-Ferreira, A., Climaco, M. C., and Mendes-Faia, A. (2001). The role of non-*Saccharomyces* species in releasing glycosidic bound fraction of grape aroma components – a preliminary study. J. Appl. Microbiol. 91, 67–71.

36. Mendoza, L. M., Manca de Nadra, M. C., and Farías, M. E. (2007). Kinetics and metabolic behavior of a composite culture of Kloeckera apiculata and *Saccharomyces cerevisiae* wine related strains. Biotechnol. Lett. 29, 1057–1063.

37. Millet, V., and Lonvaud-Funel, A. (2000). The viable but non-cultivable state of wine microorganisms during storage. Lett. Appl. Microbiol. 30, 136–141.

38. Mora, J., Barbas, J. I., and Mulet, A. (1990). Growth of yeast species during the fermentation of musts inoculated with Kluyveromyces thermotolerans and *Saccharomyces cerevisiae*. Am. J. Enol. Vitic. 41, 156–159.

39. Moreira, N., Mendes, F., Guedes de Pinho, P., Hogg, T., and Vasconcelos, I. (2008). Heavy sulphur compounds, higher alcohols and esters production profile of Hanseniaspora uvarum and Hanseniaspora guilliermondii grown as pure and mixed cultures in grape must. Int. J. Food Microbiol. 124, 231–238.

40. Nissen, P., Nielsen, D., and Arneborg, N. (2003). *Saccharomyces cerevisiae* cells at high concentrations cause early growth arrest of non-*Saccharomyces* yeasts in mixed cultures by a cell-cell contact-mediated mechanism. Yeast 20, 331–341.

41. Nurgel, C., Erten, H., Canbas, A., Cabaroglu, T., and Selli, S. (2002). Influence of *Saccharomyces cerevisiae* strains on fermentation and flavor compounds of white wines made from cv. Emir grown in Central Anatolia, Turkey. J. Ind. Microbiol. Biotechnol. 29, 28–33.

42. Pallmann, C. L., Brown, J. A., Olineka, T. L., Cocolin, L., Mills, D. A., and Bisson, L. F. (2001). Use of WL medium to profile native flora fermentations. Am. J. Enol. Vitic. 52, 198–203.

43. Pérez-Nevado, F., Albergaria, H., Hogg, T., and Girio, F. (2006). Cellular death of two non-*Saccharomyces* wine-related yeasts during mixed fermentation with *Saccharomyces cerevisiae*. Int. J. Food Microbiol. 108, 336–345.

44. Pretorius, I. S. (2000). Tailoring wine yeast for the new millennium: novel approaches to the ancient art of winemaking. Yeast 16, 675–729.

45. Rantsiou, K., Dolci, P., Giacosa, S., Torchio, F., Tofalo, R., Torriani, S., Suzzi, G., Rolle, L., and Cocolin, L. (2012) Candida zemplinina can reduce acetic acid production by *Saccharomyces cerevisiae* in sweet wine fermentations. Appl. Environ. Microbiol. 78, 1987–1994.

46. Raspor, P., Milek, D. M., Polanc, J., Mozina, S. S., and Cadez, N. (2006). Yeasts isolated from three varieties of grapes cultivated in different locations of the Dolenjska vine-growing region, Slovenia. Int. J. Food Microbiol. 109, 97–102.

47. Rojas, V., Gil, J., Piñaga, F., and Manzanares, P. (2003). Acetate ester formation in wine by mixed cultures in laboratory fermentations. Int. J. Food Microbiol. 86, 181–188.

48. Romano, P., Fiore, C., Paraggio, M., Caruso, M., and Capece, A. (2003). Function of yeasts species and strains in wine flavour. Int. J. Food Microbiol. 86, 169–180.

49. Romano, P., Suzzi, G., Comi, G., and Zironi, R. (1992). Higher alcohol and acetic acid production by apiculate wine yeasts. J. Appl. Bacteriol. 73, 126–130.

50. Schmitt, M. J., and Breinig, F. (2002). The viral killer system in yeast: from molecular biology to application. FEMS Microbiol. Rev. 26, 257–276.

51. Sipiczki, M., Ciani, M., and Csoma, H. (2005). Taxonomic reclassification of Candida stellata DBVPG 3827. Folia Microbiol. (Praha) 50, 494–498.

52. Smit, A. Y., du Toit, W. J., and du Toit, M. (2008). Biogenic amines in wine: understanding the headache. South Afr. J. Enol. Vitic. 29, 109–127.

53. Soufleros, E., Barrios, M. L., and Bertrand, A. (1998). Correlation between the content of biogenic amines and other wine compounds. Am. J. Enol. Vitic. 49, 266–269.

54. Strauss, M. L. A., Jolly, N. P., Lambrechts, M. G., and van Rensburg, P. (2001). Screening for the production of extracellular hydrolytic enzymes by non-*Saccharomyces* wine yeasts. J. Appl. Microbiol. 91, 182–190.

55. Tofalo, R., Chaves-López, C., Di Fabio, F., Schirone, M., Felis, G. E., Torriani, S., Paparella, A., and Suzzi, G. (2009). Molecular identification and osmotolerant profile of wine yeasts that ferment a high sugar grape must. Int. J. Food Microbiol. 130, 179–187.

56. Tofalo, R., Schirone, M., Telera, G. C., Manetta, A. C., Corsetti, A., and Suzzi, G. (2011). Influence of organic viticulture on non-*Saccharomyces* wine yeast populations. Ann. Microbiol. 61, 57–66.

57. Tofalo, R., Schirone, M., Torriani, S., Rantsiou, K., Cocolin, L., Perpetuini, G., and Suzzi, G. (2012). Diversity of Candida zemplinina strains from grapes and Italian wines. Food Microbiol. 29, 18–26.

58. Tofalo, R., Torriani, S., Chaves-López, C., Martuscelli, M., Paparella, A., and Suzzi, G. (2007). A survey of *Saccharomyces* populations associated with wine fermentations from the Apulia region (South Italy). Ann. Microbiol. 57, 545–552.

59. Toro, M. E., and Vazquez, F. (2002). Fermentation behavior of controlled mixed and sequential cultures of Candida cantarellii and *Saccharomyces cerevisiae* wine yeasts. World J. Microbiol. Biotechnol. 18, 347–354.

60. Torrea, D., and Ancín, C. (2002). Content of biogenic amines in a Chardonnay wine obtained through spontaneous and inoculated fermentation. J. Agric. Food Chem. 50, 4895–4899.

61. Trioli, G., and Hofmann, U. (2009). ORWINE: Code of Good Organic Viticulture and Wine-Making. Oppenheim: ECOVIN-Federal Association of Organic Wine-Producer.

62. Viana, F., Gil, J. V., Genovés, S., Vallés, S., and Manzanares, P. (2008). Rational selection of non-*Saccharomyces* wine yeasts for mixed starters based on ester formation and enological traits. Food Microbiol. 25, 778–785.

63. Zironi, R., Romano, P., Suzzi, G., Battistuta, F., and Comi, G. (1993). Volatile metabolites produced in wine by mixed and sequential cultures of Hanseniaspora guilliermondii or Kloeckera apiculata and *Saccharomyces cerevisiae*. Biotechnol. Lett. 15, 235–238.

64. Zohre, D. E., and Erten, H. (2002). The influence of Kloeckera apiculata and Candida pulcherrima yeasts on wine fermentation. Process. Biochem. 38, 319–324.

65. Zott, K., Claisse, O., Lucas, P., Coulon, J., Lonvaud-Funel, A., and Masneuf-Pomarede, I. (2010). Characterization of the yeast ecosystem in grape must and wine using real-timePCR. Food Microbiol. 27, 559–567.

There are two tables that are not available in this version of the article. To view this additional information, please use the citation on the first page of this chapter.

CHAPTER 4

WINERY WASTEWATER TREATMENT: EVALUATION OF THE AIR MICRO-BUBBLE BIOREACTOR PERFORMANCE

MARGARIDA OLIVEIRA AND ELIZABETH DUARTE

4.1 INTRODUCTION

The wine sector has faced increasing pressure in order to fulfill the legal environmental requirements, maintaining a competitive position in a global market. The rising costs associated have stimulated the sector to seek sustainable management's strategies, focussing on controlling the demand for water and improving its supply. These can be accomplished by defining the best practical techniques, using technological means (Best Available Technologies) (Duarte et al., 2004). Some EU Directives were implemented concerning water protection and management. These included in particular the Framework Directive in the field of water policy and environmental legislation about specific uses of water and discharges of substances. The disposal of the untreated waste from the wine sector is considered an environmental risk, causing salination and eutrophication of water resources; waterlogging and anaerobiosis and loss of soil struc-

Margarida Oliveira and Elizabeth Duarte (2011). Winery Wastewater Treatment - Evaluation of the Air Micro-Bubble Bioreactor Performance, Mass Transfer - Advanced Aspects, Dr. Hironori Naka-jima (Ed.), ISBN: 978-953-307-636-2, InTech, DOI: 10.5772/20805. © The Authors. Available from: http://www.intechopen.com/books/mass-transfer-advanced-aspects/winery-wastewater-treatment-evaluation-of-the-air-micro-bubble-bioreactor-performance. Licensed under a Creative Commons Attribution 3.0 Unported License, http://creativecommons.org/licenses/by/3.0/.

ture with increased vulnerability to erosion (Schoor, 2005). The winery wastewater is seasonally produced and is generated mainly as the result of cleaning practices in winery, such as washing operations during crushing and pressing grapes, rinsing of fermentations tanks, barrels washing, bottling and purges from the cooling process. As a consequence of the working period and the winemaking technologies, volumes and pollution loads greatly vary over the year. Each winery is also unique in wastewater generation, highly variable, 0.8 to 14 L per litre of wine (Schoor, 2005; Moletta, 2009). Consequently, the treatment system must be versatile to face the loading regimen and stream fluctuation. During the peak season (vintage), the winery wastewater has a very high loading of solids and soluble organic contaminant, but after this period, contaminant load decreases substantially. The high concentration of ethanol and sugars in winery wastewater justifies often the choice of a biological treatment (Bolzonella & Rosso, 2007). But the different wine processing method of each winery generates wastewater with specific properties, causing the impossibility to meet a general agreement on the most suitable cost-effective alternative for biological treatment of this wastewater.

Several winery wastewater treatments are available, but the development of alternative technologies is essential to increase their efficiency and to decrease the investment and exploration costs (Coetzee et al., 2004). So criteria should be considered in the selection of the adequate technology, such as maximization of removal efficiency, flexibility in order to deal with variable concentration and loads, moderate capital cost, easy to operate and maintain, small footprint, ability to meet discharge requirements for winery wastewater and also low sludge production. On the other hand, small producer with relatively modest financial capacity are interested in simple treatment systems with low maintenance and manpower requirements (Andreottola et al., 2009).

Most treatment systems have been designed with large oxidation tanks and oversizing the aeration system to deal with the peak load with a very high oxygen demand, during the vintage period. As a result, wastewater treatment plants are quite large and difficult to manage. One of the most promising technologies appears to be the vertical reactors characterised by high oxygen mass transfer improving the biological conversion capacity. To optimise the mass transfer, a highly efficient Venturi injector coupled with multiplier nozzles were developed (AirJection®), in order to increase the treatment efficiency.

The main goals of the present paper are the comparison of different biological treatment systems, in particular fixed and suspended biomass, operating under aerobic conditions. Since the accurate design of the bioreactor is dependent on many operational parameters, aspects related to hydraulic retention time; oxygen mass transfer and contact time, energetic costs; sludge settling and production; response time during startup, flexibility and treated wastewater reuse, in crop irrigation, with the aim of closing the water cycle in the wine sector, will be addressed. A new treatment system will be presented as a case study, an air micro-bubble bioreactor (AMBB), that will highlight the advantages and constraints on its performance at bench-scale and full-scale, in order to fulfill the gaps associated with the implemented winery wastewater treatment systems. The data presented was collected during four years monitoring plan and used to develop a tool to support the selection of the best available technology. The present study will also contribute to the implementation of an integrated strategy for sustainable production in the wine sector, based on a modular and flexible technology that will facilitate compliance with environmental regulations and potential reuse for crop irrigation. This approach will contribute to the development of a bio-based economy in the wine sector that should be integrated in a Green Innovation Economy Cycle.

4.2 COMPARISON OF DIFFERENT BIOLOGICAL TREATMENT SYSTEMS

4.2.1 BIOLOGICAL TREATMENTS IN WINERY WASTEWATER

Several treatment systems, both physico-chemical and biological, have been assayed to reduce the organic load of the winery wastewater. Some of these technologies are based on membrane bioreactors (MBRs), sequencing batch reactor (SBR), upflow anaerobic sludge blanket (UASB), anaerobic sequencing batch reactor (ASBR) and jet loop reactors (JLR). However, most of these methods have some characteristics in common: they are relatively expensive, they are not applicable in all situations, and they are not always able to deal with fluctuations in the hydraulic and pollution load. In order to overcome some of these problems, research efforts have been made towards the development of novel bioreactors as alternatives or to improve, the above-mentioned conventional methods. Although the high organic load of this wastewater would recommend the application of an anaerobic treatment for removing its polluting content, several problems have been found in the application of anaerobic processes due to its seasonal nature, its variable volumes and compositions and the difficulties in the monitoring and process control by specialized personnel (Malandra et al., 2003). The anaerobic treatments such as UASB and ASBR have successfully been used to treat a variety of effluents including those from wineries. The chemical oxygen demand (COD) removal efficiency is greater than 90%, but a specific microbial community is required. However, normally after this reactor there is an aerobic post-treatment, to return the treated water to environment (Moletta 2005) in addition, the process control needs specialized personnel (Malandra et al., 2003).

The MBRs are very compact systems and offer an alternative to conventional activated sludge processes. The COD efficiency achieved is above 97%. The electricity consumption and the operating life of the membranes are higher than those associated with traditional activated sludge systems (Artiga et al., 2005), what may constitute a constraint to its application.

The subsurface-flow constructed wetland is described as suitable for treating these wastewaters, but frequently wineries do not have available area for setting up such plants (Grismer et al., 2003). However, phytotoxicity bioassays carried out with *Phragmites, Juncus* and *Schoenoplectus* at different wastewater dilutions showed that at greater than 25% wastewater concentration all the macrophytes died (Arienzo et al., 2009a). Nevertheless, the same authors showed that this system when combined with a previously sedimentation/ aerobic process could be used for small wineries located in rural areas, achieving 72% of COD removal rate (Arienzo et al., 2009b).

The wastewater treatment of small wineries (less than 15,000 hL of wine per year) can be also performed using a SBR, fed once a day. The SBR system is a modified design of conventional activated sludge process and it has been widely used in industrial wastewater treatment. The COD removal efficiency is between 86-99% (Torrijos & Moletta, 1997). The on-line monitoring of dissolved oxygen concentration appeared as a good indicator of the progress in the COD biodegradation (Andreottola et al., 2002). Some modifications have been done in order to improve the reactor performance. The opportunity of combining the advantages of the SBR with fixed biomass was investigated (Andreottola & et al., 2002). This system permits the treatment of high organic loads, 6.3 kg COD m^{-3} d^{-1} with high biofilm grown (4-5 kg TSS m^{-3}), allowing the reduction of the required volume for biological treatment and avoiding bulking problems. However, the degradation of organic matter present in a winery wastewater sometimes require the addition of extra nutrients, to balance the C/ N/ P ratio and some oxygen efficiency transfer problems were detected when higher organic loads were applied (Lopez-Palau & Mata-Alvarez, 2009).

The fixed bed biofilm reactor or the air-bubble column bioreactor using self-adapted microbial population either free or immobilized can achieve 90% of COD removal (Petruccioli et al., 2000). In order to overcome the energetic costs associated with the aeration systems, a Venturi injector was used in the JLRs. This system achieves COD removal efficiency near 90%. Though, the high shear stress applied on the Venturi influence the composition of the microbial population (Petruccioli et al., 2000; Eusébio et al., 2005) leading to settling sludge problems. A similar technology that utilizes also a Venturi injector is the AMBB. This technology is very promising because it consists in a vertical reactor with good oxygen transfer and high biological conversion capacity. To optimise the mass transfer, a highly efficient Venturi injector coupled with multiplier nozzles was patented (AirJection®) and was applied in a lagoon system (Meyer et al., 2004) and in a vertical reactor (Oliveira et al., 2009), at pilot scale, to treat winery wastewater with a treatment efficiency of 90 %.

The maintenance and enhancement of a biological reactor is highly dependent on the microbial population that changes with time and winery activity (Jourjon et al., 2005). A deep understanding on the microbial population involved in the process is crucial to address any strategy for treatment system management (Tandoi et al., 2006). Although some researchers have been developed (Eusébio et al., 2004; Eusébio et al., 2005; Jourjon et al., 2005), the understanding of the microflora dynamics inside the bioreactor will be of utmost importance for the treatment system optimisation. Moreover, in the aerobic bioreactors the microorganisms are dependent on aeration oxygen supply. Knowledge of mass transfer coefficients between the different phases together with reaction dynamics is utmost importance to design gas-liquid-solid reactor and to predict the microbial metabolism pathway.

4.2.2 OPTIMIZATION OF OPERATIONAL PARAMETERS IN AEROBIC REACTORS

The optimization of operational parameters in bioprocesses is based essentially on reducing the volume and footprint, oxygen mass transfer and contact time, energetic costs, sludge settling and production and response time during start-up, while maintain a high removal efficiency of organic matter. The SBR system has been widely applied to organic carbon removal in municipal and industrial wastewater treatments, as this system presents different advantages such as space reduction and the ability to make operational changes, during the treatment cycle.

In the SBR system the sludge settlement occur in the same tank as oxidation, so in order to optimize the sludge settling time, the formation of granules could be performed based on feast-famine periods (Lopez-Palau et al., 2009). The start-up were performed with the increasing of the COD loading (2.7-20 kg COD. m^{-3} day^{-1}) in order to reach the feast period. After ten days of operation, the first aggregates were observed. But, the use of a high organic load promotes microbial growth and the reactor reached solids concentration of around 6 g VSS L^{-1}. Consequently, some problems of aeration appeared, and the air supply had to be increased from 13.5 L min^{-1} to 20 L min^{-1}. This study showed that is possible to cultivate aerobic granular sludge in SBR, improving the sludge settleability. Nevertheless, the aeration must be proportional to the COD load.

The combination of the SBR with fixed biomass SBBR (Sequencing Batch Biofilm Reactor) to treat winery wastewater was studied by Andreottola et al. (2002) and revealed the possibility of treating higher organic loads without increasing the required treatment volume, as the biomass grown on plastic media. However, this type of reactor needs a separated settler, as the biomass settlement worsens in the presence of the plastic material. In order to optimize the energetic costs and the SBBR performance, a strategy based on dissolved oxygen (DO) monitoring was developed.

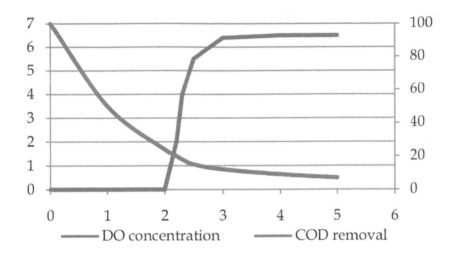

FIGURE 1: DO concentration and COD dynamic during a typical SBBR cycle

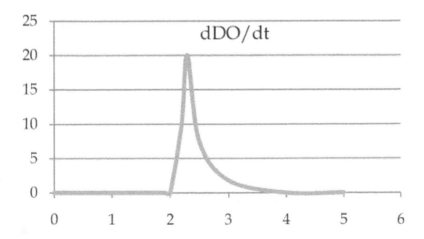

FIGURE 2: Time derivative of DO concentration

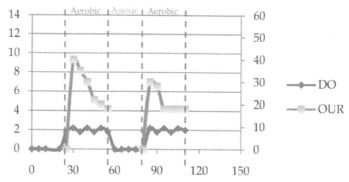

Adapted from Puig *et al.* (2006).

FIGURE 3: DO and OUR evolution during aerobic and anoxic phases

During the assays the DO control was the key to the COD removal, because this treatment was carried out with constant aeration (up to 4.5 hours). When the treatment started the COD decreases as the DO concentration is maintained at low levels (Figure 1). Once the microbial activity decreases by diminishing the organic load, the DO concentration begins to rise until reaching a plateau. At this stage, the process is complete and the cycle can be stopped. The end of each cycle can be calculated based on the first derivative function of the DO concentration vs time (Figure 2). With this strategy it was possible to reduce the hydraulic retention time in about three times, which has allowed the treatment of a higher flow with a similar effluent quality.

Another approach based on dissolved oxygen control was carried out to optimize a SBR cycle for total organic carbon and ammonia removal (Puig et al., 2006). In this treatment the aerobic phases of the SBR cycle were initially operated using an On/ Off dissolved oxygen control strategy.

The cycle was divided in reaction phase, under aerobic and anoxic conditions, settling and discharge. During the aerobic phase a fixed DO set-point of 2.0 mg DO L-1 was applied, as a simple On/ Off control. The system optimization was based on pH, DO and OUR evolution. This strategy allows the detection of the ammonia valley in the pH profile and

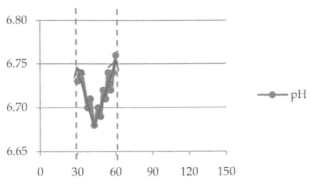

Adapted from Puig *et al.* (2006).

FIGURE 4: Detection of the ammonia valley in the pH evolution during aerobic phase

also the end of nitrification, through the OUR outline (Figure 3 and Figure 4). The analysis of the OUR profile shows a plateau in the OUR value, in the end of the aerobic phase, which may indicate that the microbial populations are under endogenous conditions and that organic matter and ammonia has been completed degraded. In fact, one of the most important aspects in many biological systems is the aeration supply.

The wastewater treatment is one of these processes that require proper aeration to maintain the growth of the microorganisms responsible for the biodegradation of the organic matter. Most wastewater treatments are aerobic and are carried out in aqueous medium containing inorganic salts and organic substances which can give viscosity to the broth, showing a non-Newtonian behavior. In bioprocessing it is very important to ensure an adequate oxygen distribution to the gas stream and to the fermentation broth. Some of the systems used to supply the oxygen are sparging, free-jet flow and bubbling column, among others. Also the different nozzle geometry, the liquid phase properties, the jet length and diameter influences the oxygen distribution to the system that in many cases is a limiting factor to the success of the treatment process. In this sense, it is important to estimate the mass transfer characteristics in order to predict the kinetic growth reaction constant, and control and optimize the aerobic fermentation processes

(Choi et al., 1996; Fakeeha et al., 1999; Tojabas & Garcia-Calvo, 2000; Garcia-Ochoa & Gomez, 2009). The volumetric mass transfer coefficient, kLa, is the parameter that characterizes the gas-liquid mass transfer in bioreactors. However, this value can vary substantially from those obtained for the oxygen absorption in water or in simple aqueous solutions, and in static systems with invariable composition of the liquid media along time. The transfer rate is very influenced by the nature of pollutants present in the wastewater, for example glucose increases the medium viscosity causing a decrease in the kLa value while the low foam surfactants enhances this value (Fakeeha et al., 1999; Tojabas & Garcia-Calvo, 2000). Thus, it is necessary to know the composition of the fermentative broth, at least some of the major compounds, to understand the effect of combination of different pollutants for proper design and operation of aerobic process. Many strategies have been proposed to determine the volumetric mass

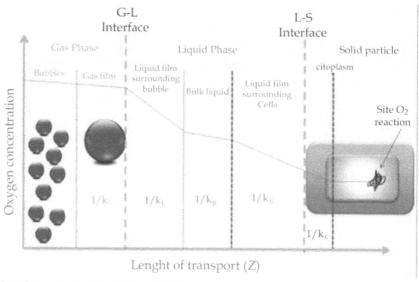

Adapted from Garcia-Ochoa & Gomez (2009)

FIGURE 5: Steps and resistances for oxygen transfer from gas bubble to cell, in three phases reactors

transfer coefficient, empirical equations and also theoretical prediction, most of them developed for bubble columns and airlifts (Garcia-Ochoa & Gomez, 2009).

The bioprocesses involves simultaneous transport and biochemical re-actions, the oxygen is transferred from a rising gas bubble to the liquid phase and then to the place of oxidative phosphorylation within the cell, considered as a solid particle. The steps related to this mass transfer pro-cesses can be represented according to the film theory model for mass transfer, which describes the flux through the film based on a driving force (Figure 5). The oxygen mass transfer rate per unit of reactor volume is ob-tained by a solute mass balance for the liquid phase (Fakeeha et al., 1999):

$$OTR = k_L a \times (C^* - C_L) \tag{1}$$

As k_L and a are difficult to measure separately, usually the $k_L a$ is evalu-ated together and this parameter is identified as the volumetric mass trans-fer coefficient that characterizes the gasliquid mass transfer. The driving force is the gradient between the oxygen concentration at the interface and in the bulk liquid. This gradient varies with the solubility and microbial activity. Also, the gas solubility depends on temperature, pressure, con-centration and type of salts present in the system.

In bioreactors it is essential to determine the experimental kLa to set the aeration efficiency and to quantify the effects of the operating variables on the dissolved oxygen supply. To select the appropriated method, some factors should be taken into account, such as aeration system; bioreactor type and its mechanical design; the composition of the fermentation broth and the possible effect of the microorganisms (Xu et al., 2010).

The mass balance for the dissolved oxygen in the well-mixed liquid phase can be established as (Garcia-Ochoa & Gomez, 2009; Irizar et al., 2009):

$$(dC/dt) = OTR - OUR \tag{2}$$

Where dC/ dt is the accumulation oxygen rate in the liquid phase, OTR is the oxygen transfer rate from the gas to the liquid phase, described by equation (1) and OUR is the oxygen uptake rate by microorganisms. The methods that can be applied for the oxygen transfer rate measures can be classified depending on whether the measurement is done in the absence of microorganisms or with dead cells or in the presence of biomass that consumes oxygen at the time of measurement. When biochemical reactions do not take place, OUR=0, then the equation (2) can be simplified to:

$$(dC/dt) = k_L a \times (C^* - C_L) \tag{3}$$

The dynamic method used to measure the $k_L a$ value is based on the dissolved oxygen consumption and supply. In this method the change in the dissolved oxygen concentration is analyzed supplying air until the oxygen saturation concentration in the liquid phase is reached. The oxygen decreasing is then recorded as a function of time. Under these conditions the equation (2) can be expressed as equation (4), but, after the decreasing phase, the oxygen is again supplied and the equation (2) can be written as equation (5). In these cases the $k_L a$ values can be determined from the slope of the ln $f(C_L)$ vs time.

$$\ln (C_{L0}/C_L) = - k_L a \times t \tag{4}$$

$$\ln (1 - C_L/C^*) = - k_L a \times t \tag{5}$$

Furthermore $k_L a$ is usually expressed at standard conditions of temperature and pressure, 20°C, 1atm (equation 6).

$$k_L a_{20} = k_L a_T \times 1.024^{20-T} \tag{6}$$

The determination of the oxygen uptake rate OUR can also be carried out using a dynamic method which measures the respiratory activity of microorganisms that grow in the bioreactor. When the air supply is switching off, the dissolved oxygen concentration will decrease at a rate equal to oxygen consumption due to the microorganisms respiration rate. In this situation the OUR is determined from the slope of the plot of dissolved oxygen concentration vs time. The biomass concentration should be known in order to determined the specific oxygen uptake rate (SOUR).

Another important parameter in the aerobic reactors optimization is the sludge settling and production. The large amount of excess sludge generated during activated sludge process is estimated to cost about 40-60 % of the operating cost (Chen et al., 2001). This sludge contains volatile solids and retains about 95% of water resulting in a large volume of residual solids produced. The biological sludge production in conventional wastewater treatment plants can be minimized using different strategies (Pérez-Elvira et al. 2006), such as endogenous metabolism and maintenance metabolism. In this last approach part of energy source is used for maintaining living functions, in this phase the substrate consumption is not used for cellular synthesis. In the endogenous metabolism part of cellular components is oxidized to produce the required energy for maintenance functions, which leads to a decrease in the biomass production. The objective is to reach a natural balance between biomass growth and decay rates. The oxic-settling-anoxic activated sludge process, considered as a sludge feast/ famine treatment, is based on alternating exposure of sludge to oxic and anoxic environments. This working principle stimulate catabolic activity and make catabolism dissociate from anabolism. The sludge famine is related to an exposure of the settled sludge to anoxic conditions where the substrate concentration is low. Under these stressful conditions microorganisms are starving which may lead to a depletion of cell energy or nutrients storage. The sludge feasting means that fasted microorganisms return to an oxic environment with enough nutrients. As a consequence, the microorganisms growth may be limited by energy uncoupling (Chen et al., 2001).

FIGURE 6: Winery activities

4.2.3 DIAGNOSIS PROCESS

The selection of the most appropriate technology for the winery wastewater treatment is a difficult step that should be done after a diagnosis process. A proper diagnosis should conduct a survey report that includes all the information required for decision-makers. Regarding the production process, it should address all activities associated with it: vintage, racking and bottling (Figure 6). The knowledge of materials and supplies, as well as byproducts generated during the process is essential in diagnosis. The water uses and water consumption are critical, both in terms of quantity or quality. The survey of sewers in the farm unit, particularly if the drainage system is separated or combined, and the points of wastewater discharge should also be covered. The wastewater flows should be evaluated through the installation of flow meters. The different streams of wastewater generated must be quantified in order to make an assessment, as rigorous as possible.

The water consumption in two Portuguese wineries, one small and one medium size are quite different, with regard to quantity. However, the distribution of water consumption has a similar behavior throughout the year (Figure 7 and Figure 8). The data presented show that most water (60%-80%) is consumed in the vintage period that last about a month. So, the collection of water consumption associated with the physicochemical characterization of the wastewater is essential for the proper sizing of any

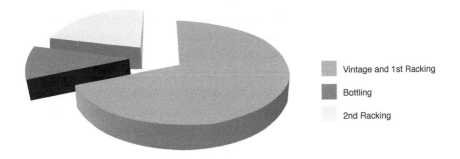

FIGURE 7: Distribution of water consumption during the global period of processing at a medium dimension winery

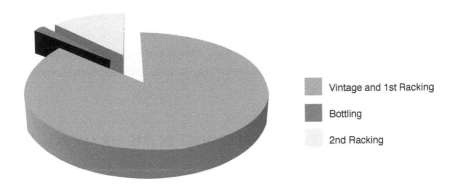

FIGURE 8: Distribution of water consumption during the global period of processing at a small dimension winery

treatment system. In addition, it is possible to understand the need for flexibility of the treatment system, because the system should allow good removal yields, during the vintage period, but has also to remain in operation during the rest of the year even at low loads. In small wineries often there is a minor stream of wastewater during several months, which may lead to bioreactor inefficient performance. To overcome this situation a feast/ famine strategy may be a challenge for future research at a full-scale.

The physicochemical characterization assessment is carried out by determining specific parameters such as pH, electrical conductivity, dissolved oxygen, chemical oxygen demand, biochemical oxygen demand, total phosphorus, total nitrogen, total solids, suspended solids, total polyphenol compounds, anionic surfactants. In order to evaluate the fate of treated wastewater, it is also important to know the winery surroundings, in particular the existence of a sewage, the irrigation area, the type of structures and available areas, among others.

In wineries that intend to reuse the treated wastewater for irrigation, other concerns should be considered. The domestic wastewater flow containing high concentration of pathogenic microorganisms should not be mixed with the industrial wastewater stream. This flow should be treated separately or discharged in the sewage. This decision is extremely important, since the wastewater from winery operations does not contain pathogenic microorganisms. Thus, this separation reduces the costs of wastewater treatment and monitoring, which are associated with the disinfection process.

4.3 WINERY WASTEWATER TREATMENT IN THE AIR MICRO-BUBBLE BIOREACTOR

4.3.1 WASTEWATER CHARACTERIZATION

The winery wastewater was collected, during four years, from three wineries of different sizes and characteristics and located in different Regions of Portugal. The Casa Agrícola Quinta da Casa Boa, located at Runa, Lisboa Region, producing only red wines, has a small/ medium dimension with a production capacity of 200,000 L. The Catapereiro, located at Alcochete, Tejo Region, produces both white and red wines, has a medium dimension with a production capacity of 1 000,000 L of wine. The Herdade da Mingorra, located at Beja, Alentejo Region, has a medium dimension with a production capacity of 1 000,000 L of wine (Figure 9 to Figure 14).

Composite samples of the winery wastewater, representative of each phase of the process, were taken and maintained at 4°C. A set of major key parameters were defined and analysed, according to Standard Methods for the Examination of Water and Wastewater (1998), in order to assess the winery wastewater pollutant charge: pH, conductivity, chemical oxygen demand (COD), biochemical oxygen demand (BOD), total suspended solids (TSS), volatile suspended solids (VSS), polyphenols, anionic surfactants, Na, K, Mg and Ca. The winery wastewater flows were evaluated from water consumption. With this propose the wineries installed general water counters to be daily read and register.

4.3.2 BIOREACTOR SET-UP

The Air Micro-Bubble Bioreactor (AMBB) with a total volume of 15 dm³ consists of a cylindrical bioreactor, equipped with a circulated pump and a settler (Figure 15). The aeration was conducted during the wastewater recirculation by a high efficiency Venturi injector (HEVI) in conjunction with mass transfer multiplier nozzles (MTM). The MTM nozzles discharge the air/ water mixture from the HEVI into the bottom of the bioreactor (Figure 14). The AMBB is equipped with an air flow meter and a monitoring probe (HANNA Instruments) able to on-line monitor pH, DO and temperature. Figure 16 shows a schematic overview of the bioreactor.

4.3.3 BIOREACTOR START-UP AND OPERATING CONDITIONS

Several trails performed with the AMBB, under batch conditions were carried out during 15 days. The reactor was inoculated with 15 dm³ of fresh winery wastewater, from the vintage period and with 0.15 dm³ of acclimated biomass, obtained during the treatment of winery wastewater, in the previously year. Samples from the mixed liquor were daily taken for physico-chemical characterisation. The aerated flow was 2 dm³ min⁻¹. The operating temperature was 20-30°C. The recirculation of the mixed

FIGURE 9: Global view at Mingorra winery

FIGURE 10: Mingorra winery unit

FIGURE 11: Vintage at Quinta da Casa Boa

FIGURE 12: 1st Racking at Quinta da Casa Boa

FIGURE 13: Wastewater treatment system at Catapereiro

FIGURE 14: Wastewater treatment system at Mingorra

FIGURE 15: The air micro-bubble bioreactor filled with clean water

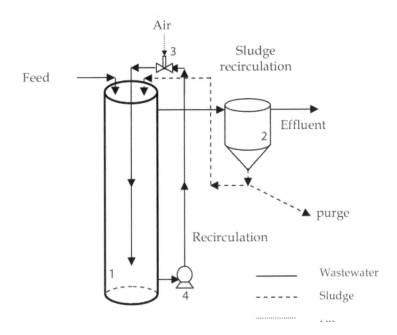

FIGURE 16: Flow diagram of the air micro-bubble bioreactor. 1- Bioreactor; 2- Settler; 3- Venturi injector; 4- Recirculation pump

liquor started with 20 min hour[-1], with a flow of 40 dm^3 min[-1] and then was changed to 5 min hour[-1].

4.3.4 SEED GERMINATION BIOASSAYS

Germination bioassays were performed following Fuentes et al. (2004), by using cress *Lepidium sativum* L. seeds, to evaluate the suitability of the treated wastewater in relation to crop irrigation and expressed as Germination Index. The treated wastewater and two dilutions in distilled water (25%, 50% v/ v) were tested (Oliveira et al., 2009).

4.4 RESULTS AND DISCUSSION

4.4.1 WASTEWATER ASSESSMENT

During the studying period, samples of winery wastewater were taken for laboratory characterization to evaluate their pollutant charge (Table 1). The values of pH ranged from 4 to 8, being this variation mostly dependent on the labor period. The electric conductivity of the wastewater showed no relevant variation in the different sampling periods and the range of registered values is not considered as inhibiting biomass growth.

The highest values of COD were reached during the vintage period, followed by the first racking. These results are in accordance to those previously reported by other authors (Petruccioli et al., 2002). As expected, the highest values of biodegradability (BOD_5/ COD) were achieved during the vintage period, due to the high concentration of simple molecules, easily metabolized (sugars and ethanol) by microorganisms (Duarte et al., 2004).

Concerning TS and TSS parameters, the results reveal a high variability during the vinification period. Moreover, the TS are significantly higher than TSS, which means that these wastewaters contain, mostly, dissolved organic pollutant charge. However, during 2nd racking the TSS concentration reach the maximum value derived from the presence of tartrate. These solids are often problematic due to the high phenolic load adsorbed.

Although polyphenols and anionic surfactants are important pollutants, it is not expected that they could influence the organic load, since they are present in low concentration. Nevertheless, after the wastewater treatment some compounds known as recalcitrant may remain in the treated effluent, such as the polyphenols that are responsible for colour and the residual COD, this can also be observed by the low biodegradability ratio presented in Figure 17.

Moreover, this type of wastewater has very low levels of nutrients that are essential to microbial growth. For this reason, it is often required the addition of nutrients to guarantee the process of cellular synthesis. Alternatively, it is possible to change some practices at the winery in order to balance this ratio (Oliveira & Duarte, 2010).

The assessment of the water consumption is another key parameter for the successful of the winery wastewater treatment. In one of the monitored wineries the water consumption was evaluated throughout the operation period for two consecutive years. Internal management strategies were implemented to increase efficient water use, such as cleaning methods that aim the water reuse (closed-loop) pressure washing machines, among others. These simple changes showed a saving in water consumption of about 40%.

4.4.2 AMBB TREATMENT

In this type of seasonal industry, the treatment system must be able to treat the wastewater produced in the vintage period. For this reason, many reactors have an appropriate volume for this stage but over dimensioned during the rest of the year. On the other hand, the high organic load of these wastewaters may promote the excessive growth of biomass, that requires an increase in the air supply (López-Palau et al., 2009) and creates problems of sludge generation and disposal.

The adopted strategy in this study is based on sludge reduction, as the production of excess sludge from the wastewater treatment plant is considered one of the serious problems encountered in the aerobic treatments (Liu & Tay, 2001). In this study, an aerobic step alternated with an anoxic one was adopted as a strategy.

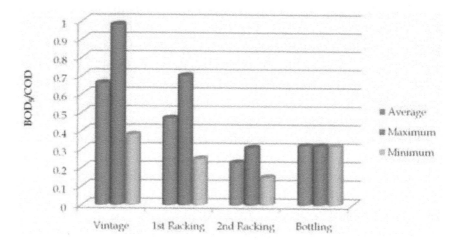

FIGURE 17: Biodegradability indicators of the winery wastewater, in different labour periods

FIGURE 18: Inoculation of the AMBB with fresh winery wastewater

FIGURE 19: AMBB in the beginning of the treatment

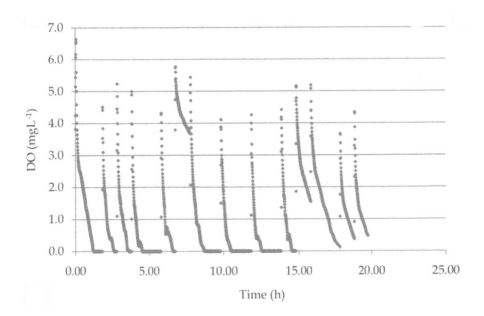

FIGURE 21: Evolution on pH and DO concentration in the AMBB

The bioreactor AMBB was tested in different phases of the wine process and it started in the vintage period (Figure 18-19). The biomass inoculated in this assay was already acclimated to winery wastewater and was maintained in aerobic/ anoxic conditions, with insufficient substrate. During the AMBB operation the microorganisms grow as suspended biomass but also as biofilm adsorbed to the reactor walls.

The evolution of COD concentration, biomass and dissolved oxygen was followed. Regarding the biomass evolution, a typical growth curve for batch cultivation was achieved (Figure 20). This curve does not show a lag phase, since biomass was already adapted. The recirculation of the mixed liquor was 20 min hour^{-1}.

The COD of the winery wastewater ranged between 4.0-8.0 kg COD m^{-3} but the efficiency was similar for each batch, about 90.0±4.3%, after 6 days of operation. This period is related to the biomass exponential phase. The maximum efficiency obtained (98.6±0.4%) was achieved after 15 days of treatment. These results are comparable with those reported by Beltran de Herédia et al. (2005), where they achieve 75% of COD reduction, after 3 days of treatment.

In order to minimize the sludge production and the energetic costs during the recirculation of the mixed liquor, the aeration time was reduced. During this assay dissolved oxygen, pH, COD and biomass was evaluated.

Concerning the DO concentration the Figure 21 illustrate the dynamic change of this parameter in the AMBB. During the air supplying, the DO increases until it reaches saturation. The period of time required to reach saturation is directly related to the oxygen transfer rate. The estimation of OTR under different operational conditions has a relevant role to predict the metabolic pathway for microbial growth in aerobic treatments. So, this approach could be interesting for studying the influence of operational conditions on volumetric mass transfer coefficient.

A dynamic method was used to determine the volumetric mass transfer coefficient, k_La (Table 2). The kLa values were calculated by solving the Equation 2, during the aeration phase and considering that the gas flow and OUR were both constant. In these cases the slope of the ln f(DO) vs time allows the determination of the oxygen transfer parameter (Figure 22). The k_La values were corrected to 20 ° C, according to equation 6.

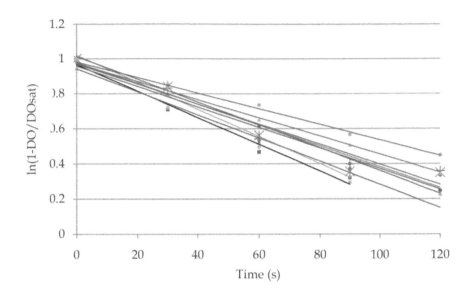

FIGURE 22: Experimental determination of kLa based on DO concentration in the AMBB

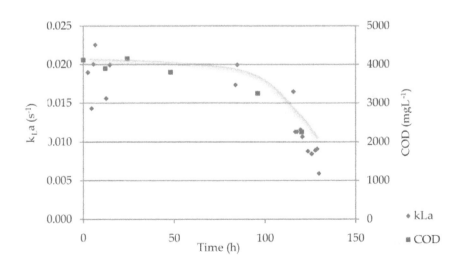

FIGURE 23: kLa and COD dynamics during wastewater treatment in the AMBB

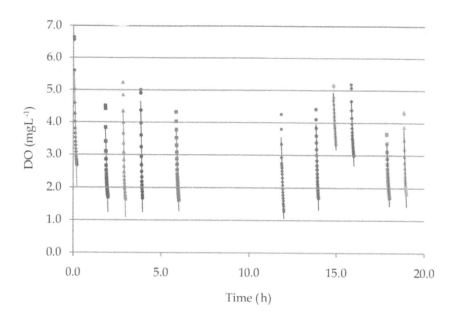

FIGURE 24: Trendlines adjustment on DO concentration depletion to determine OUR in the AMBB

The results show a decrease in the $k_L a$ value during the treatment period (Figure 23). Many factors could influence $k_L a$, including air flow rate, air pressure, temperature, vessel geometry and fluid characteristics. All parameters were kept constant throughout the treatment, except the wastewater composition that varies during the treatment period. More readily biodegradable compounds such as sugars and ethanol are firstly assimilated by microorganisms; the more complex substrates are only degraded at a later stage. Previous studies indicate that the composition of the fermentation broth influences the oxygen mass transfer, such as glucose that can decrease the $k_L a$, by increasing the viscosity of the medium but on the other surfactants increases this value (Fakeeha et al., 1999). In fact, is practically impossible to determine the exact composition of wastewater, but the compounds mentioned above are always present in this type of wastewater. This decrease in $k_L a$ value means that some of the existing compounds in wastewater optimize the oxygen mass transfer during the initial phase of treatment. Indeed, even without being quantified, in all trials it was found that the size of the bubble formed, increased throughout the treatment period, which is in agreement with the obtained results. Moreover, it is interesting to observe that the $k_L a$ decline follows the degradation kinetics of organic matter, expressed as COD, which corroborate the obtained results. The $k_L a$ values obtained in these assays are in the same range that of values achieved by other authors in full-scale aeration tank equipped with fine bubble diffusers and jet loop reactor (Fakeeha et al., 1999; Fayolle et al., 2010). In addition, the respirometric activity of microorganisms which are actively growing in the bioreactor can also be measured based on this dynamic method. When the gas supply to the bioreactor is turned off, the DO concentration decreases at a rate equal to oxygen consumption by the respiration process. In this situation the OUR can be calculated from the slope of the DO vs time (Figure 24).

The specific oxygen uptake rate (SOUR) or respiration rate is expressed as milligrams of oxygen consumed per gram of volatile suspended solids per hour. The high SOUR values obtained (Table 2), indicate a high organic load to the existing suspended solids in the mixed liquor (MLSS).

The SOUR measurements throughout the wastewater treatment showed an initial increase in the SOUR values until reaching a plateau. Figure 25 shows that after an adaptation period to the treatment system there is a removal of the organic load, expressed as COD rate corresponding to the increment of SOUR rate. This high SOUR rate is due to the high activity of the microbial population to oxidise substrates. These values may be induced by an increased energy requirement stimulated by a famine period, during sludge acclimatisation. The feast/ famine phenomenon has been reported by several authors as a strategy on sludge production (Chen et al., 2001; Ramakrishna & Viraraghavan, 2005; López- Palau et al., 2009). A similar behaviour was found by Chen et al. (2001) during the study of feast/ famine growth on activated sludge cultures previously subjected to a famine treatment. This study also indicates that the COD removal ability of the fasted culture is higher than the non-fasted culture.

In winery wastewater treatments systems the period prior to vintage is a non-productive period, without wastewater generation. In this sense, the existing biomass in the treatment system is subjected to a famine treatment. Moreover, during harvest the wastewater production has the highest flow rates and organic loadings. According to Chen et al., 2001 after a famine period the microorganisms are starved and the substrate utilization rate increases. A treatment system based on this management model seems to be a good approach for winery wastewaters, with the additional advantage of keeping the low amount of sludge. The cause of sludge reduction in this process is not clearly known but the absence of oxygen reduces the growth of strictly aerobic populations and stimulates the facultative bacteria (unpublished results), which have lower specific growth rates. In this sense, as the dominant population is constituted of slow growers that may explain the low sludge yield production. Furthermore, the produced sludge shows low SVI values indicative of easy sludge settling.

The strategy based on low aeration time alternating with anoxic periods allows the treatment of the winery wastewater with lower sludge production but with lower efficiency. In fact, the MLSS achieved in this

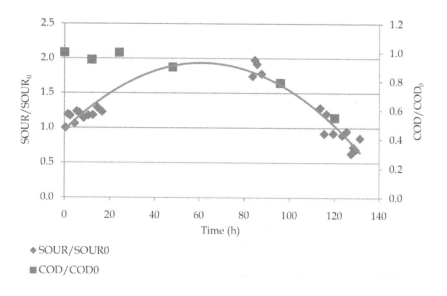

FIGURE 25: Evolution of SOUR and COD rates during the batch treatment

batch treatment, 1.2 g/ L was lower compared with the initial assay. In the management of a wastewater treatment of this nature is necessary to establish a compromise between operating costs and final quality of the treated wastewater, taking into account the final destination and the legal requirements.

In order to evaluate its suitability to be used in crop irrigation the treated wastewater from the AMBB batch assays was physico-chemical characterized. All the analyzed parameters except one were in agreement with EU and Portuguese Legislation (Directive 2000/ 60/ EC, DL n° 236/ 98) for irrigation use (Table 3). Of particular concern was the sodium adsorption ratio (SAR), the proportion of sodium to calcium and magnesium, which was higher than the permitted parametric value. Probably, some strategies can be applied in wineries in order to reduce the problem. Nevertheless, the treated wastewater, normally, is used in irrigation systems to supplement the irrigation water, as an economic additional water supply. Also, seed germination assays carried out with Lepidium sativum were developed for evaluating the effects of water contaminants on germination and seedling growth. The adequacy of the treated wastewater for crop irrigation was evaluated with direct toxicity bioassays, by using cress seeds as indicator No significant differences (P=0.05) between batch experiments were registered on germination index (GI). As the cress bioassay is a standard procedure to evaluate the behaviour of crops to water contaminants, data (previously published) evidence the suitability of treated wastewater in relation to crop irrigation, thus minimizing water consumption (Oliveira et al., 2009).

4.5 CONCLUSION

In this type of seasonal industry, the treatment system must be able to treat the wastewater produced in all labour period. A vertical reactor coupled with highly efficient Venturi injector and multiplier nozzles was used for winery wastewater treatment. Regarding mass transfer parameters, the estimation of OTR could be interesting for studying the influence of operational conditions on volumetric mass transfer coefficient. The results showed a decrease in the kLa value during the treatment period. This decrease in kLa value may evidence that some of the existing compounds in wastewater optimize the oxygen mass transfer during the initial phase of treatment.

The SOUR measurements throughout the wastewater treatment showed high values, which could indicate a high organic load to the existing suspended solids in the mixed liquor. This high SOUR rate is due to the high activity of the microbial population to oxidise substrates, in the begging of the treatment.

The implemented strategy, where an aerobic step alternated with an anoxic one was adopted, showed to be a good approach to minimize the sludge production and to reduce energy waste. However, further studies should be conducted in order to better understand the effect of winery wastewater composition in mass transfer coefficients. The use of two consecutive stages of treatment might improve the performance of this technology because it allows higher flexibility. The result of feast / famine treatment in sludge should also be exploited, as it is of interest in this type of seasonal industries. Moreover, the treated wastewater revealed its suitability to be integrated in the irrigation systems as confirmed by direct toxicity bioassays. This study is expected to contribute to the implementation of an efficient wastewater treatment, intending the preservation of the water resource, the reduction of the wastewater sludge production and the energy safe.

REFERENCES

1. Andreottola, G., Foladori P., Ragazzi, M. & Villa, R. (2002). Treatment of winery wastewater in sequencing batch bofilm reactor. Water Science and Technology Vol. 45, No.12, pp. 347–354, ISSN 0273-1223.

2. Andreottola, G., Foladori P. & Ziglio, G. (2009). Biological treatment of winery wastewater: an overview. Water Science and Technology Vol. 60, No.5, pp. 1117–1125, ISSN 0273- 1223.

3. A.P.H.A., A.W.W.A., W.E.F. (1998) Standard methods for the examination of water and wastewater, 20a Ed., USA.

4. Arienzo, M., Christen, E.W., Quayle, W (2009a). Phytotoxicity testing of winery wastewater for constructed wetland treatment. Journal of Hazardous Materials, Vol. 169, No. 1, pp. 94-99, ISSN 0304-3894.

5. Arienzo, M., Christen, E.W., Quayle, W, Di Stefano, N. (2009b) Development of a Low-Cost Wastewater Treatment System for Small-Scale Wineries. Water Environment Research, Vol. 81, No. 3, pp. 233-241, ISSN 1061-4303.
6. Artiga,P.,Ficara,E.,Malpei,F.,Garrido,J.M. & Méndez,R.,2005. Treatment of two industrial wastewaters in a submerged membrane bioreactor. Desalination, Vol. 179, pp. 161-169, ISSN 0011-9164.
7. Beltran de Herédia J., Torregrosa J., Dominguez J.R. and Partido E. (2005). Degradation of wine distillery wastewaters by the combination of aerobic biological treatment with chemical by Fenton's reagent. Water Science and Technology, Vol. 51, No. 1, pp. 167-174, ISSN 0273-1223.
8. Bolzonella, D. & Rosso, D. (2007). Winery wastewater characterisation and biological treatment options. Water Science and Technology, Vol. 56, No. 2, pp. 79–87, ISSN 0273-1223.
9. Chen, G., Yip, W., Mo, H., Liu, Y. (2001). Effect of sludge fasting/ feasting on growth of activated sludge cultures. Water Research. Vol. 35, No. 4, pp.1029-1037, ISSN: 0043-1354.
10. Choi, K.H., Christi, Y., Moo-Young, M. (1996). Comparative evaluation of hydrodynamic and gas-liquid mass transfer characteristics in bubble column and airlift slurry reactors. The Chemical Engineering Journal, Vol. 62, pp. 223-229. ISSN 0923-0467.
11. Coetzee G., Malandra, L., Wolfaardt, G.M. & Viljoen-Bloom, M., (2004). Dynamics of microbial biofilm in a rotating biological contactor for the treatment of winery effluent. Water SA, Vol. 30, No. 3, pp. 407-412.
12. Duarte E., Reis I.B., Martins M.O. (2004). Proceedings of 3rd Internat. Special. Conf. Sustainable Vitic. Winery Wastes Management, Barcelona 23-30.
13. Eusébio A., Mateus M., Baeta-Hall L., Almeida-Vara E., Duarte J.C. (2004). Microbial characterization of activated sludge in Jet-loop bioreactor treating winery wastewater. Journal of Industrial Microbiology and Biothechonogy. Vol. 31, 29-34, ISSN 1476-5535.
14. Eusébio A., Mateus M., Baeta-Hall L., Almeida-Vara E., Duarte J.C. (2005). Microflora evaluation of two agro-industrial effluents treated by the JACTO jet-loop type reactor system, Water Science and Technology. Vol. 51, 107-112, ISSN 0273-1223.
15. Fakeeha, A.H., Jibril, B.Y., Ibrahim, G., Abasaeed, A.E. (1999). Medium effects on oxygen mass transfer in plunging jet loop reactor with a downcomer. Chemical Engineering and Processing, Vol. 38, pp. 259-265, ISSN: 0255-2701.
16. Fayolle, Y., Gillot, S., Cockx, A., Bensimhon, L., Roustan, M., Heduit, A. (2010). In situ characterization of local hydrodynamic parameters in closed-loop aeration tanks. Chemical Engineering Journal, Vol. 158, pp. 207-212, ISSN: 1385-8947
17. Fuentes A., Lloréns M., Sáez J., Aguilar M.I., Ortuño J.F. and Meseguer V.F. (2004). Phytotoxicity and heavy metals speciation of stabilized sewage sludges. Journal of Hazardous Materials, Vol. 108 No. 3, pp. 161–169, ISSN: 0304-3894.
18. Garcia-Ochoa, F. & Gomez,E. (2009). Bioreactor scale-up and oxygen transfer rate in microbial processes: An overview. Biotechnology Advances, Vol. 27, pp.153-176, ISSN: 0734-9750.
19. Grismer M.E., Carr M.A., Shepherd H.L. (2003). Evaluation and constructed wetland treatment performance for winery wastewater. Water Environment Research, Vol. 75, pp. 412-421, ISSN 1061-4303.

20. Irizar, I., Zambrano, J.A., Montoya, D., Gracia, M., Garcia, R. (2009). Online monitoring of OUR, KLa and OTE indicators: pratical implementation in full-sacele industial WWTPs. Water Science and Technology Vol. 60, No. 2, pp. 459-466, ISSN 0273-1223.

21. Jourjon, F., Khaldi, S., Reveillere, M., Thibault, C., Poulard, A., Chretien, P., Bednar, J. (2005). Microbiological characterization of winery effluents: an inventory of the sites for different treatment systems , Water Science and Technology. Vol. 51, 19-26, ISSN 0273-1223.

22. Liu, Y. & Tay, J.H. (2001). Strategy for minimization of excess sludge production from activated sludge process. Biotechnology Advances, 19, 97-107, ISSN: 0734-9750.

23. López-Palau, S. Dosta, J. & Mata-Alvarez, J. (2009). Start-up of aerobic granular sequencing batch reactor for the treatment of winery wastewater. . Water Science and Technology Vol. 60, No. 4, pp. 1089-1095, ISSN 0273-1223.

24. Malandra, L. Wolfaardt, G., Zietsman, A., Viljoen-Bloom, (2003). Microbiology of biological contactor for winery wastewater treatment. Water Research, Vol. 37, pp. 4125-4134, ISSN: 0043-1354.

25. Meyer R.M., Mazzei A.L. and Mullin J.R. (2004). Aerobic treatment of winery wastewater utilizing new aeration technology. In: Proceedings of the 3rd International Specialised Conference on Sustainable Viticulture and Winery Wastes Management, Barcelona, 353-355.

26. Moletta, R. (2009). Biological treatment of wineries and distillery wastewater. Proceedings of 5th International Specialized Conference on Sustainable Viticulture Winery Wastes and Ecological Impact Management. pp. 389-398, ISBN 978-88-8443-284-1, Trento and Verona, 30 Mar-3 Apr.

27. Moletta R. (2005). Winery and distillery wastewater treatment by anaerobic digestion. Water Science and Technology Vol. 51, No.1, pp. 137–144, ISSN 0273-1223.

28. Oliveira, M., Queda, C., Duarte, E. (2009). Aerobic treatment of winery wastewater aiming the water reuse. Water Science and Technology Vol.Vol. 60, No. 5, pp.1217-1223. ISSN 0273-1223.

29. Oliveira, M. & Duarte, E. (2010). Guidelines for the management of winery wastewaters. Proceedings of 14th Ramiran International Conference on Treatment and Use of Organic Residues in Agriculture: Challenges and Opportunities Towards Sustainable Management, Lisboa, 12-15 Sept.

30. Pérez-Elvira, S. I., Diez, P.N., Fdz-Polanco, F. (2006). Sludge minimisation technologies. Reviews in Environmental Science and Biotechnology, Vol. 5, pp. 375-398, ISSN: 1569-1705.

31. Petruccioli M., Duarte J.C., Eusébio A. and Federici F. (2002). Aerobic treatment of winery wastewater using a jet-loop activated sludge reactor. Process Biochemistry, Vol. 37, No. 8, pp. 821-829, ISSN: 1359-5113.

32. Petruccioli M., Duarte J.C. and Federici F. (2000). High-rate aerobic treatment of winery wastewater using bioreactors with free and immobilized activated sludge. Journal of Bioscience and Bioengineering ,Vol. 90, No. 4, 381-386, ISSN: 1389-1723.

33. Puig, S., Corominas, L., Traore, A., Colomer, J., Balaguer, M.D., Colprim, J. (2006). An online optimization of a SBR cycle for carbon and nitrogen removal based on

on-line pH and OUR: the role of dissolved oxygen control. . Water Science and Technology Vol. 53, No.4-5, pp. 171–178, ISSN 0273-1223.

34. Racault Y. and Stricker A.E.(2004). Combining membrane filtration and aerated storage: assessment of two full scale processes treating winery effluents. In: Proceedings of the 3rd International Specialised Conference on Sustainable Viticulture and Winery Wastes Management, Barcelona, 105-112.

35. Ramakrishna, D.M. & Viraraghavan, T. 2005. Strategies for sludge minimization in activated sludge process – a review. Fresenius Environmental Bulletin, Vo. 14, No.1, pp. 2-12, ISSN: 1018-4619.

36. Tojabas, M., Garcia-Calvo, E. (2000). Comparison of experimental methods for determination of the volumetric mass transfer coefficient in fermentation processes. Heat and Mass Transfer, Vol. 36, pp. 201-207.

37. Tandoi, V., Jenjins, D., Wanner, J. (2006). Activated sludge separation problems. Scientific Technical Report 16, 216 pp., ISBN: 1900222841.

38. Torrijos, M. & Moletta, R. (1997). Winery wastewater depollution by sequencing batch reactor. Water Science and Technology Vol. 35, No.1, pp. 249–257, ISSN 0273-1223.

39. van Schoor, L.H. (2005). Guidelines for the management of wastewater and solid waste at existing wineries. Winetech, 35 pp.

40. Xu, Y. Zhou, J., Qu, Y., Yang, H., Liu, Z. (2010). Dynamics and oxygen transfer of a noval vertical tubular biological reactor for wastewater treatment. Chemical Engineering Journal Vol. 156, pp. 92–97, ISSN 0273-1223.

There are several tables and one figure that are not available in this version of the article. To view this additional information, please use the citation on the first page of this chapter.

CHAPTER 5

THE IMPORTANCE OF CONSIDERING PRODUCT LOSS RATES IN LIFE CYCLE ASSESSMENT: THE EXAMPLE OF CLOSURE SYSTEMS FOR BOTTLED WINE

ANNA KOUNINA, ELISA TATTI, SEBASTIEN HUMBERT, RICHARD PFISTER, AMANDA PIKE, JEAN-FRANÇOIS MÉNARD, YVES LOERINCIK, AND OLIVIER JOLLIET

5.1 INTRODUCTION

5.1.1 CONTEXT AND OBJECTIVES

The environmental impacts of wine have been assessed by several studies [1,2,3,4,5,6,7,8,9,10,11,12,13,14,15]. Some of the studies consider the environmental impacts of different closure systems and offer comparative conclusions [16,17]. However, different types of closures, such as natural cork stoppers, synthetic stoppers or screw caps, have different properties, offering different levels of product protection and consequently presenting more or less risk for wine losses. To date, the influence of the closure type on the overall environmental impacts of bottled wine, taking losses into account, has not been studied in a life-cycle assessment context.

In this study, wine loss refers to loss occurring when the consumer does not consume the wine contained in the bottle and disposes of it because of taste alteration, which is caused by inadequate product protection rendering the wine unpalatable to a knowledgeable consumer. When studying different bottled wine systems, it is essential to ensure that the compared systems are functionally equivalent and therefore, there is a need to consider the differences in wine loss rates due to better or worse product protection due to the closure. This study addresses this need and analyses the influence of closures on the environmental impacts of bottled wine, accounting for differences in loss rates for two selected closure systems: cork stoppers and screw caps, as shown in Figure 1.

FIGURE 1: Cork stopper and screw cap closures.

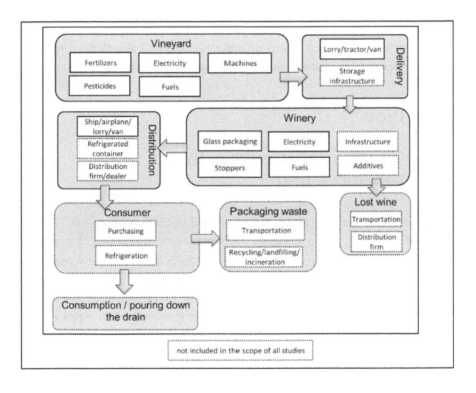

FIGURE 2: System for bottled wine and its closure system including the different life cycle stages.

The objective of this study is to discuss the implications of loss rates in terms of environmental performance depending on the closure type of the bottled wine system. Considering the entire system contributes to providing reliable and fair results for sustainable product design, branding and marketing. We quantify both the life-cycle environmental profile of the closure itself and the impacts related to the differences in wine loss rates for each closure. Environmental impacts associated with wine losses are derived from the environmental impact of typical bottled wine with a defined quality, based on a literature review. The difference in wine alteration rates was estimated based on interviews with experts performed at universities and laboratories as well as a literature review [18,19].

5.2 METHOD

5.2.1 CLOSURE SYSTEMS CONSIDERED AND FUNCTIONAL UNIT

The most wide-spread closure systems used in the world market [20] are (1) cork stoppers (approximately 60% of world market), (2) synthetic stoppers (approximately 30% of world market), and (3) screw caps (approximately 10% of world market). "Crown stoppers" are used only for low-end wines or during the manufacturing of effervescent wines and therefore their market share is negligible. In this study, two types of closures were chosen (cork stoppers and screw caps) and their impacts were compared based on their material properties as well as their respective wine loss rates.

The system studied relates to the functional unit "a 750 mL bottle of drinkable wine" and includes the following main components:

- wine bottle production
- 750 mL of wine
- wine closure production
- wine loss associated with the closure systems
- wine poured down the drain

Other studies considering a delivered bottled wine as a functional unit did not consider if bottled wine quality affected whether or not the consumer would actually drink the wine. Quantities of bottled wine produced were recalculated for all closure types based on the equivalent functional unit of 750 mL of drinkable bottled wine, thus including wine losses. This functional unit allows for the comparison of studies evaluating the life cycle impact of bottled wine production, as well as the consideration of wine loss rates due to different types of closures.

5.2.2 SYSTEM BOUNDARIES AND BOTTLED WINE PRODUCTION

The studied system includes the raw material extraction, vineyard operations, delivery, closure and packaging manufacturing, winery operations, distribution, consumer purchasing and packaging end of life (Figure 2). The wine poured down the drain is included in the system, and the related model is described in Section 2.4.

A literature review has been performed on studies assessing the life cycle impact of wine production, independent of the closure type [1,2,3,4,5,6,7,8,9,10]. Only the vineyard operations, delivery, winery and distribution stages are taken into account in all these studies. Management of the waste generated by the vineyard, retail functioning and infrastructure, reclosing opened bottles, vineyard and winery overheads (offices, employee commuting and business travels, marketing, advertising, and other administrative functions), as well as wine glass use and washing are not taken into account in all studies and are therefore not considered in this literature review comparison. Part of the ethanol consumed by humans and converted to CO_2 by human metabolism represents biogenic CO_2 and has not been considered in this system.

The reviewed studies present different indicators to assess the environmental impact of wine. In order to ensure the highest level of comparability among the analyzed references, this literature review focuses on the following indicators that were assessed in most of the reference studies:

1. Global warming over 100 years (in kg CO_2eq)

2. Non-renewable primary energy use (in MJ)
3. Atmospheric acidification (in g H$^+$ eq)
4. Photo-oxidant formation (in kg ethylene eq)
5. Eutrophication of surface water (in kg PO$_4^{3-}$ eq)

These indicators are included in the International Reference Life Cycle Data System (ILCD) framework for impact categories for characterization modeling at midpoint [21]. Although the studies consulted provide an essential basis for gaining an understanding of the overall environmental impact of wine production, they also have some important limitations. They do not all have the same scope or system boundaries, nor do they use the same impact assessment method. However, as the impact results on common indicators have the same order of magnitude, the differences in scope and system boundaries are not significant enough to prevent them being used for this case study. For other applications, care must be taken if results are to be used to determine the absolute impact of a bottle of wine.

A detailed description of the results of the literature review can be found in the supporting information Section S1.

5.2.3 WINE CLOSURE PRODUCTION

The impacts of closure systems (including production of raw materials, transport of raw materials, manufacturing of closures, transport of closures, bottling and end-of-life) are directly taken from the Corticeira study [17].

5.2.4 WINE LOSS ASSOCIATED WITH THE CLOSURE SYSTEMS

Physical losses which occur during filling, packaging, and transporting (breaking bottles, spilling wine) and which are independent of closure types are not taken in account in this study, which focuses solely on the influences of different closure types. Physical failures during all life-cycle stages which result in wine alteration, e.g., poor storing of closures, poor calibration of the bottling machinery or transport in inappropriate conditions (such as inappropriate temperature, exposure to the sun or humidity, etc.), are considered and summarized in supporting information in Table S2, where the fraction of wine taste alteration per stage has been estimated.

Wine taste alteration occurs when the contents of a bottle acquires an altered taste, either due to inadequate protection allowing for oxidation of the content or via a reaction with the closure material, such that the consumer decides not to drink it and disposes of it. For this study, the level of wine loss is estimated by multiplying the following two parameters:

1. Percentage of bottles with an altered taste (Section 2.4.1). This parameter depends on the type of closure as well as wine conservation failures during its life cycle.
2. Percentage of the bottles with an altered taste that are actually thrown away by the consumer (Section 2.4.2 on consumer behavior). This parameter depends on the consumer, their expertise and sensitivity to recognize the altered taste (due to the wide range of alteration intensity).

5.2.4.1 TASTE ALTERATIONS CONSIDERED AND FRACTION OF BOTTLES WITH ALTERED TASTES

The taste alterations considered in this study are those discovered by the consumer or sommelier. Taste alterations can occur throughout the entire life cycle of bottled wine, from storage in a wine cellar to the consumer's glass. The following closure system failures can be distinguished at the consumer level:

1. an undesirable odor due to failures at different life cycle stages (e.g., incorrect transport or storage in inappropriate locations before and after the bottling process);
2. an unfavorable taste due to conservation problems (e.g., being stored too long);
3. a "corked" taste.

Several "corked tastes" are possible and appear with differing frequencies. A description of each is as follows:

1. A "true corked taste"—or "goût de bouchon"—is due to the presence of the fungus *Armillaria mellea* [22] and is characterized by a powerful taste which makes the wine undrinkable; this is also known as "yellow stain";
2. A "taste of cork" occurs when wine elements are affected by the cork. The wine is considered altered when it is too intense;
3. A "mould taste" is caused by microorganisms on the cork (e.g., *Aspergillus* and *Penicillium*), which affect the organoleptic character of wine [23,24];
4. A "false corked taste" is due to 2,4,6-trichloroanisol (TCA), which is synthesized by the fungi *Penicillium* from chlorophenols, which can enter cork during its production and/or storage.

Mold tastes and false corked tastes are the most frequent wine alterations.

Cork stoppers: Wine taste alteration due to cork stoppers is primarily due to the presence of fungus which results in the "corked taste" [19]. Wine taste alteration can also occur during transport, storage, cork machine adjustment as well as when the cork's lifespan is exceeded. Undesirable odors are often erroneously attributed to the "true cork taste", which causes considerable uncertainty in the percentage of bottles affected by this phenomenon [25]. In addition, there are three types of cork stoppers (chipboard cork, natural cork, and cork treated with supercritical carbon dioxide) [26,27,28] that can cause variable taste alteration, making it yet more difficult to determine with certainty that wines possess the "true corked taste".

Given these uncertainties, taste alteration rates due to cork stoppers can be only approximately estimated using expert judgment. Seven interviews were carried out and complemented with two literature sources to serve as primary data for determining the respective fractions of bottled wine with altered tastes for the two types of closures studied. Experts from cellar quality control consultancies, laboratories and various institutes specialized in bottling or cellar quality control were interviewed. Their estimations for wine alteration rates are listed in the supporting information S2. The share of bottles with cork stoppers having an altered taste is estimated to be between 2% and 5%, where the lower and higher values correspond to the medians of the limits provided by the experts and literature sources.

This range reflects the variability of existing cork stopper quality, their treatment, and the sensitivity of the type of wine to taste alteration.

Screw caps: Unpleasant odors due to the reduction of wine can occur if the choice of the screw cap liner is not appropriate. Several types of liners exist with various degrees of permeability and are normally tailored to the specific application. Based on discussions with the same companies specialized in bottling or cellar quality control, the taste alteration rate with the screw caps is reduced by a factor 5 to 10 compared to the cork stoppers (see detailed in supporting information S2). The share of bottles with screw caps having an altered taste is therefore estimated to be lower than that for cork stoppers by a factor of 7.5 times (i.e., between 0.3% and 0.7% of bottles). However, we have included in the analysis a broad loss range (from a factor 5 to 10 times less than cork stoppers) and considered a triangular distribution (5 times lower (maximum), 7.5 times (mean) and 10 times (minimum)) for the modeling in order to consider the uncertainty resulting from all input parameters variability.

An estimation of the life cycle stage at which the failure originated is provided in the supporting information S3. We acknowledge the limitations of data availability on the respective taste alteration rates due to different closure types. However, the order of magnitude of the relative wine taste alteration rates for both types of closures estimated by the expert sources consulted are considered sufficiently robust for them to be used in this study. For this reason, we have carried out a sensitivity study looking at the lower and upper range of effective losses considered.

5.2.4.2 CONSUMER BEHAVIOR

In this study, it is assumed that altered wine odors or tastes result in different consumer behaviors which are guided by subjective taste perception, wine quality expectation and whether the consumer is a sommelier or a regular consumer (recognizing that the latter will also have a range of behaviors). These behaviors could include:

1. Drinking the bottle (considered unlikely for a strongly "corked" bottled wine)

2. Disposal of the wine down the drain (or returning it to the store before eventual disposal at the store) and opening a new bottle. Thus, this particular behavior increases the amount of wine needed to fulfill the functional unit

In the reference scenario, the consumer behavior is estimated based on a combination of literature data and expert judgment. The European average for consumers who would drink the bottle without considering wine to be "altered" is based on the MIS Trend study [29] which estimated that 23% of Swiss consumers considering themselves as not knowledgeable about wine. The other 77% would replace a bottle of wine with an altered taste with a new bottle. With these insights, an estimated effective wine loss between 1.5 and 3.8% for cork stoppers and between 0.21 and 0.51% for screw caps were determined.

To complement this reference scenario and to test the sensitivity of impacts to changes in wine taste alteration and replacement rates, we carried out a sensitivity study looking at the impact variations of the cork and screw cap closures as a function of these rates. Based on this analysis, environmental preference can be deduced for replacement rates based on specific consumer behavior patterns.

5.2.5 IMPACTS OF WINE POURED DOWN THE DRAIN

Two types of emissions are considered when the bottled wine with altered taste is poured in the sewage system: (1) direct emissions, i.e., methane emissions which occur from the sewage and wastewater treatment plant and (2) indirect emissions, i.e., from functioning of the wastewater treatment plant and its related electricity and chemicals production. Direct CO_2 emissions are not considered since this represents biogenic CO_2 that has been fixed earlier during wine production (not considered in the literature sources for bottled wine production), leading to a zero net emission over the wine life cycle.

Direct emissions: Wine that is poured down the drain consists of organic compounds, mainly ethanol (estimated at 10% of the wine by volume), which is broken down into carbon dioxide (CO_2), and, under anaerobic conditions, into methane (CH_4) in the sewage system and the anaerobic sludge digester. The calculation details for the initial amount of carbon (31 g) poured into the sewage system are specified in supporting information S4. Figure 3 shows the emissions of methane and carbon dioxide due to wine poured down the drain and entering the wastewater treatment plant (in % of carbon originally present in the wine, based on SimpleTreat 3.1 model [30]). Input and output parameters are summarized in supporting information S4. The Dutch Ministry of Housing, Spatial Planning and the Environment (VROM) [31,32] estimates that 0.7% of the COD mass is converted to CH_4 in a wastewater treatment plant without anaerobic sludge digestion based on the IPCC inventory guidelines [33]. This corresponds to 2.8% kg CH_4 per kg TOC, using the ratio 4 g COD/g TOC (estimation explained in the supporting information 4). Given the limitation in data availability on ethanol degradation and methane production in the sewage system, we considered the sewage and wastewater treatment system as a single system with the same physical and biological processes. Total methane emissions to air from wine poured down the drain represent 2.9E-2 kg CO_2eq per bottle as the contribution to the global warming impact category, knowing that the methane global warming potential is 25 kg CO_2eq/kg methane considering a 100-year time horizon. Carbon emissions to water after wastewater treatment produce 3.4E-4 kg PO_4^{3-}/bottle poured down the drain as the contribution to freshwater eutrophication. The latter impact score is calculated assuming an emission of 12.6% of the initial ethanol amount in the effluent (based on SimpleTreat 3.1 model), a chemical oxygen demand/total organic carbon ratio of 4, and using the IM

Indirect emissions: In addition, emissions also occur during the treatment processes. Values from ecoinvent v 2.2 [35] are used in this study and assume that an average of 750 mL of tap water is used in addition to 750 mL of bottled wine wasted. The global warming score associated with the tap water production (750 mL/bottle wasted) and the wastewater treatment plant operation (2 × 750 mL/bottle wasted) is very low, estimated at 6.3E-4 kg CO_2eq/bottle wasted.

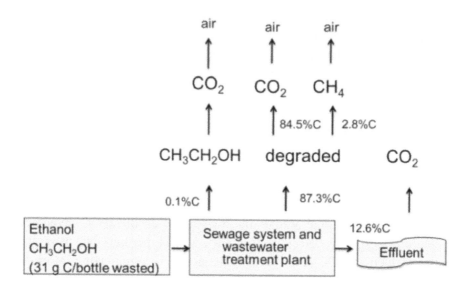

FIGURE 3: Methane and carbon dioxideemissions (associated with carbon contained in ethanol) due to wine poured down the drain and treated at the wastewater treatment plan.

5.3 RESULTS

5.3.1 IMPACT SCORES FOR WINE PRODUCTION, CLOSURES AND WINE POURED DOWN THE DRAIN

Table 1 presents the impacts of the production of 750 mL of wine, the impacts of closure production and the impacts due to 750 mL wine being poured down the drain, which were obtained from the literature review. The latter value needs to be multiplied by the loss rate in order to be re-lated to the functional unit.

TABLE 1: Impact scores for wine production, closure production and wine poured down the drain.

Impact category studied	750 ml bottled wine production [1,2,3,4,5,6,7,8,9,10]		Closure production [17]		Wine poured down the drain (recalculated based on ecoinvent v2.2)
	average	range	Cork stopper	Screw cap	For 750 mL of wine
Global warming (kg CO_2eq)	3.3	1.0-4.0	2.0E-3	3.7E-2	2.9E-2
Non-renewable energy use (MJ)	47	16-58	0.10	0.44	1.1E-3
Atmospheric acidification (g H^+ eq)	0.78	0.27-1.3	1.3E-3	8.2E-3	2.7E-4(Modeled with TRACI [36])
Photo-oxidant formation (kg ethylene eq)	1.9E-03	1.1E-3-2.3E-3	3E-6	1.4E-5	1.6E-7
Eutrophication (kg PO_4^{3-}eq)	4.5E-03	1.4E-3-7.8E-3	6E-7	7E-7	3.5E-4

The environmental profile generated by the literature review shows that the closure system represents between 0.01% and 1.1% of the total score for the produced bottled wine, depending on the indicator and the type of closure. Although minimal, the closure contribution to the overall impact appears to be largest for the global warming and atmospheric acidification impact categories. The contribution to different impact categories due to a bottle of wine poured down the drain vary between 0.002% and 7.7% (for eutrophication) of the total score for bottled wine depending on the indicator and the type of closure. Again, the latter value needs to be multiplied by the loss rate in order to be related to the functional unit.

5.3.2 INFLUENCE OF THE WINE LOSS RATE DUE TO DIFFERENT CLOSURES

The overall environmental performance of the closure in packing and preserving wine prior to its consumption combines the impacts from:

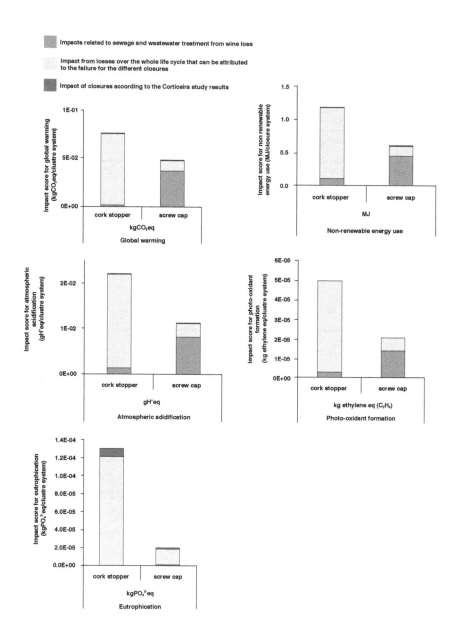

FIGURE 4: Impact over the whole life cycle associated with 750 mL drinkable wine, for a 77% replacement rate, differentiated due to losses that can be attributed to the failure of different closure systems and the impact of the closure systems themselves (excluding the non-closure-specific impacts that are equal for all scenarios).

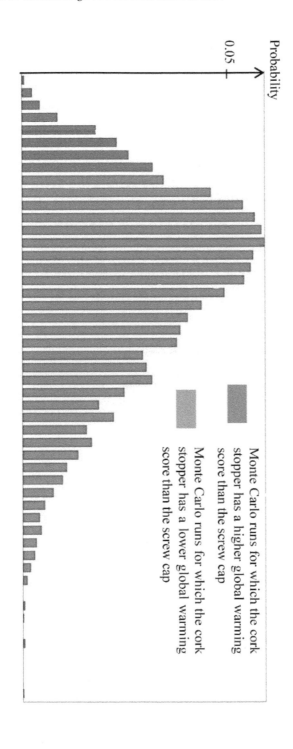

FIGURE 5: Monte Carlo analysis with 5000 runs for the cork and screw cap stopper systems for the global warming category.

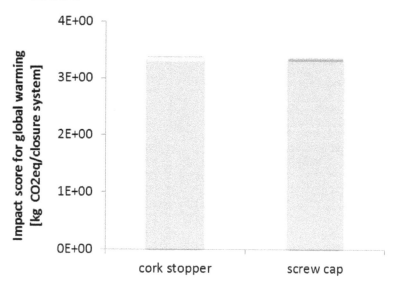

■ **Impacts related to sewage and wastewater treatment from wine loss**

Impact from losses over the whole life cycle that can be attributed to the failure for the different closures

■ **Impact of closures according to the Corticeira study results**

Average impact of a wine bottle according to the literature review

FIGURE 6: Comparison of the closure production global warming scores, including the relative loss rates concerning the entire impacts associated with bottled wine.

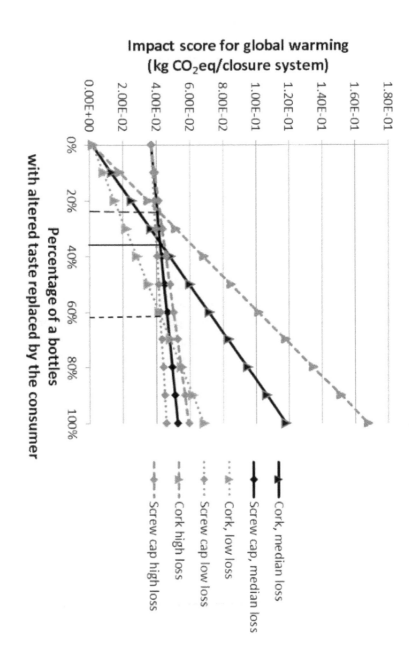

FIGURE 7: Impact score for global warming of the cork stopper production compared to the screw cap production as a function of the replacement rate

1. Losses at the consumer due to altered taste
2. Closure impacts according to the results of the Corticeira study [17]
3. Impacts related to sewage and wastewater treatment from wine loss

The reference scenario is based on the impact from wine losses when 77% of consumers replace the bottle when they encounter one with an altered taste. A sensitivity study was carried out to study the influence of replacement variation.

5.3.2.1 REFERENCE SCENARIO

Figure 4 shows the results for the reference scenario, presenting the geometric mean value of the impact from the wine loss distribution (all parameters are modeled with a statistical distribution). Throughout the whole paper, the only common functional unit is 750 mL drinkable wine. Impacts of the bottled wine are not closure-specific and are equal for both closure systems. They can therefore be excluded from Figure 4 to focus result presentation on the influence of closures and related losses. The lost bottles of wine (i.e., the extra amount of wine needed to fulfill the function of having bottled drinkable wine) can contribute anywhere from 28% of the impacts of the closures themselves (e.g., in the case of screw caps for the global warming category) to several times more than the impact of the closures themselves (e.g., in the case of cork stoppers for all impact categories). When the wine loss rate is considered, the cork stopper presents a higher score than the screw cap for all impact categories.

An uncertainty analysis was performed to assess the variability of the impact for both systems. The relative ranking for all impact categories based on a 5000-run Monte Carlo assessment was also performed. The impact of wine, cork stoppers, screw caps and replacement rates are modeled with a triangular distribution. This distribution has been selected because (1) it is defined with a most probable value as well as a 95% confidence range and (2) it can be asymmetrical. The alteration rate for the cork stopper (estimated from 2 to 5%) is modeled with a uniform distribution given there is no preferred value within this range. The alteration rate for the screw cap is estimated as 0.13 times the impact of the cork stopper (7.5 times less), with a triangular distribution between 0.1 and 0.2 times the impact of the cork stopper. The share of runs where the cork stopper system has a higher impact score than the screw cap system is specified in Table 2 for each impact category. Figure 5 displays the Monte Carlo analysis results for global warming where the cork stopper has a higher impact score than the screw cap stopper (red color) for 89.6% of the 5000 runs. More details on the uncertainty analysis parameters and results are included in the supporting information S5.

Figure 6 shows the overall contribution to the global warming impact category for bottled wine comparing the wine loss rates from the different closure systems and the closures based on the Corticeira study with the impact of the wine itself. The influence of the closure on the total impacts of bottled wine is relatively small. This explains why even a small difference in loss rates for the two types of closures can offset the impact score trend for the bottled wine and closure system originally in favor of cork stoppers. The overall impact of bottled wine shows the same trend for all impact indicators: the cumulative impact of the closure, the associated loss rate and the impacts related to sewage and wastewater treatment from wine loss, which represents less than 5% of the total impact of the whole bottled wine system.

5.3.2.2 SENSITIVITY STUDY

Since the replacement and loss rates may vary widely depending on consumer behavior, it is important to perform a sensitivity study on how these rates influence the overall impact. Figure 7 shows the global warming impact score for the cork stopper compared to the screw cap as a function of the replacement rate, for low (2% for cork stopper, 0.3% for screw caps), medium (3.5% for cork stopper, 0.5% for screw caps) and high (5% for cork stopper, 0.7% for screw caps) wine taste alteration rates. At low replacement rates, the contribution of the screw cap to the global warming impact category is higher. As the replacement rates increases, impacts due to losses for the cork stopper are more important than for the screw cap. The break-even point is at a 35% replacement rate for the average wine alteration rate of 3.5%. Thus, the product of these two rates leads to an effective loss rate of 1.2% for the cork stopper. This effective loss rate of 1.2% also corresponds to break-even points of 61% replacement for a low (2%) taste alteration rate and to 24% for a high (5%) taste alteration rate. At a higher replacement rate (e.g., the 77% rate of the reference scenario), the screw cap system clearly becomes advantageous.

TABLE 2: Monte Carlo analysis results for the reference scenario and the break-even percentages for the replacement rate and the effective losses for the sensitivity analysis, for which the total impact of the screw cap is equal to the impact of the cork stopper.

	Reference scenario	Sensitivity analysis	
Indicator	Monte Carlo analysis: % runs for which the impact score for cork stoppers is higher than for screw caps	Break-even replacement rates for which the impact of the screw cap is equal to the impact of the cork stopper (bottles with altered taste estimated as 3% for cork stoppers and as 0.5% for screw caps)	Break-even effective loss rates for which the impact of screw cap is equal to the impact of the cork stopper
Global warming (kg CO2eq) Screw cap: 0.16% of wine bottles	89.6%	35% of consumers replace the bottle	Cork stopper: 1.2% of wine bottles
Non-renewable energy use (MJ) Screw cap: 0.11% of wine bottles	98.4%	24% of consumers replace the bottle	Cork stopper: 0.83% of wine bottle
Atmospheric acidification (g H+ eq) Screw cap: 0.14% of wine bottles	96.9%	29% of consumers replace the bottle	Cork stopper: 1.0% of wine bottle
Photo-oxidant formation (kg ethylene eq) Screw cap: 0.09% of wine bottles	99.8%	19% of consumers replace the bottle	Cork stopper: 0.67% of wine bottle
Eutrophication (kg PO43-eq) Screw cap: 0.00034% of wine bottles	100%	0.07% of consumers replace the bottle	Cork stopper: 0.0026% of wine bottle

Figures and equations used for the sensitivity analysis to determine the break-even point for the other impact categories are provided in the supporting information S6. Table 2 summarizes the break-even points for replacement and effective loss rates for all impact categories. The replacement rates at the break-even points for photo-oxidant formation, non-renewable energy and atmospheric acidification are slightly lower than for global warming (19% to 29% replacement rate and 0.67% to 1.0% effective loss rate for cork stoppers). The replacement rate is very low for eutrophication (less than 0.1%). For any higher value than the break-even point, the impact of the cork stopper would exceed the impact of the screw cap.

5.3.2.3 RESULTS SUMMARY

The results of the reference scenario uncertainty as well as the sensitivity analysis are shown in Table 2.

5.4 DISCUSSION AND CONCLUSIONS

5.4.1 ACHIEVEMENTS

This study provides a better understanding of the global environmental profile of bottled wine considering the different closures used and taking into account the loss rates induced by each closure system.

1. The main conclusions of the study are as follows:
2. The different closures and associated wine losses represent less than 5% of the total life-cycle impact of bottled wine. The recently published article by Point [14] confirms that the cork stopper contribution to the total bottled wine system represents between 0.4% and 5%. Impacts related to bottled wine are composed of a several dozens of processes (Point 2008) that contribute to the global impact. While recognizing that the closure impacts are only one element of the overall system, it is only by optimizing the thousands of products and product parts that we are using in our everyday

lives that we can move towards sustainability. Reducing closure-related impact represents an easily implementable impact reduction opportunity. Other system parts should also be considered (e.g., bottle production, consumer transport) for further impact reduction opportunities.

3. The wine loss rate resulting from the type of closure and its specific properties is a key parameter to consider when assessing the impact of different wine closures and can result (especially for cork stoppers) in a higher impact than the closure itself. The reduction of wine closures impacts demonstrates the trade-off between the impact of closure and the associated losses.

4. In the case of a cork stopper, the impact of wine loss is larger than the impact of the cork stopper production itself for all examined life cycle impact categories.

5. When the impact of wine loss is considered in addition to the impact of the closure itself, the cork stopper has a higher environmental impact score than the screw cap in all impact categories, provided the effective loss rate of cork stoppers is higher than 1.2%.

In general, LCAs of packaged food consider the following:

• The functional unit used as a reference to estimate the environmental impacts of packaged food should represent an equivalent function among all types of packaging and thus must consider the different associated loss rates.

5.4.2 LIMITATIONS

The main limitations of this study include the loss rate estimation and the calculation for the impact of wine production.

Loss rate estimation: the estimated wine loss rates presented in this study are based on (1) expert opinions (i.e., not empirically measured) that reflect their perceived judgment on the percentage of bottled wine that reaches a consumer with an altered taste for each of the closures systems and (2) a sensitivity study on the influence of the consumer behavior.

Impact of wine production calculation: water use and solid waste generation may also be relevant impact indicators as they were reported in the Corticeira study. These indicators were analyzed for this study, but due to inconsistencies in reporting in the original study as well as in the studies reviewed on the impact of wine production, they are not reported here. Whilst the results for the impact of wine production reflect the state-of-the-art in life cycle assessment, the different environmental indicators are not equally covered in terms of sources, system boundaries and life cycle impact assessment methods used. Most observations and conclusions are based on—and are therefore mainly valid for—global warming and use of non-renewable energy resources, which are the most widely covered in the wine production impact assessment literature.

5.4.3 FURTHER NEEDS

To address these limitations, the following areas were identified for further exploration:

- The estimation of wine taste alteration rates induced by the cork stoppers and screw caps can be refined in future studies by sampling wines and studying their individual taste alteration rate according to key parameters, such as the type of wine, closure or cap quality, and storage conditions.
- A comprehensive survey on consumer behavior once an altered taste has been identified is needed to better quantify replacement rates and to examine which fraction of the wine is still reused for alternative purposes such as cooking.
- Several differences linked to closure functionality or closure and bottle characteristics (e.g., the additional amount of glass which is needed to accommodate a cork stopper inside the bottle) need to be analyzed and incorporated in further studies.
- The results shown in Figure 6 indicate that there is an opportunity to further improve the environmental performance of other components such as the glass bottle itself, the consumer purchasing trip, vineyard operations, winery processing and transportation, before focusing further on the wine closure system.
- In order to derive conclusions regarding the environmental superiority of closure systems that could be used in a comparative assertion, a comprehensive and full life cycle impact assessment of all indicators (midpoint and endpoint level) including the full functional equivalence must be carried out.

Regarding these limitations and needs, the present study primarily shows the clear importance of accounting for losses in packaging studies for which the impact of the content is much higher than the packaging itself. This study is not intended to be used for comparative assertions.

This study exemplifies the importance of considering the full implication of different components on overall product functionality to draw conclusions regarding environmental preferences. Sustainability in product design, branding and marketing can only be achieved through a fair and comprehensive product assessment. This conclusion is not only applicable for the food and beverage industry, but also to numerous other sectors of the economy such as the electronics industry where the reject or fail rate in the quality control step of electronic component production and resulting losses can be significant [37].

REFERENCES

1. Point, E.V. Life cycle environmental impacts of wine production and consumption in Nova Scotia, Canada. Master Thesis, Dalhousie University, Halifax, Nova Scotia, Canada, 2008.
2. Aranda, A.; Zabalza, I.; Scarpellini, S. Economic and Environmental Analysis of the Wine Bottle Production in Spain by Means of Life Cycle Assessment; Indersience: Geneva, Switzerland, 2005.
3. Ardente, F.; Beccali, G.; Cellura, M.; Marvuglia, A. POEMS: A case study of an Italian wine-producing firm. Environ. Manage. 2006, 38, 350–364, doi:10.1007/s00267-005-0103-8.
4. Colman, T.; Paster, P. Red, White and "Green": The Cost of Carbon in the Global Wine Trade; Working paper, American Association of Wine Economists, New York, NY, USA, 2007.
5. Environmental Product Declaration. Bottled Red Sparkling Wine. Validated Environmental product declaration N° S-P-00109. Available online: http://gryphon.environdec.com/data/files/6/7505/EPD%20S-P-00109%20ingl-2008-def.pdf (accessed on 15 October 2012).
6. Environmental Product Declaration, Bottled Organic Red Sparkling Wine. Validated Environmental Product Declaration N° S-P-00119. Available online: http://gryphon.environdec.com/data/files/6/7521/EPD_Fratello%20Sole_english%202008-def.pdf (accessed on 15 October 2012).
7. Niccolucci, V.; Galli, A.; Kitzes, J.; Pulselli, R.M.; Borsa, S.; Marchettini, N. Ecological footprint analysis applied to the production of two Italian wines. Agr. Ecosyst. Environ. 2008, 128, 162–166, doi:10.1016/j.agee.2008.05.015.

8. Pizzigallo, A.C.I.; Granai, C.; Borsa, S. The joint use of LCA and emergy evaluation for the analysis of two Italian wine farms. J. Environ. Manage. 2008, 86, 396–406, doi:10.1016/j.jenvman.2006.04.020.
9. Institut Français de la Vigne et du Vin. Bilan carbone: de la vigne a la bouteille. 2007. Presentation. Available online: http://www.vignevin-sudouest.com/publications/itv-colloque/documents/COLLOQUE_bilan-carbone-vigne-bouteille.pdf (accessed on 15 October 2012).
10. Gonzalez, A.; Klimchuk, A.; Martin, M. Life Cycle Assessment of Wine Production Process: Finding Relevant Process Efficiency and Comparison with Eco-Wine Production; Report, Royal Institute of Technology, Stockholm, Sweden, 2006.
11. Pattara, C.; Raggi, A.; Cichelli, A. Life cycle assessment and carbon footprint in the wine supply-chain. Environ. Manage. 2012, 49, 1247–1258, doi:10.1007/s00267-012-9844-3.
12. Gazulla, C.; Raugei, M.; Fullana-i-Palmer, P. Taking a life cycle look at crianza wine production in Spain: Where are the bottlenecks? Int. J. Life. Cycle. Ass. 2010, 15, 330–337, doi:10.1007/s11367-010-0173-6.
13. Bosco, S.; Di Bene, C.; Galli, M.; Remorini, D.; Massai, R.; Bonari, E. Greenhouse gas emissions in the agricultural phase of wine production in the Maremma rural district in Tuscany, Italy. Ital. J. Agron. 2011, 6, 93–100.
14. Point, E.; Tyedmers, P.; Naugler, C. Life cycle environmental impacts of wine production and consumption in Nova Scotia, Canada. J. Clean. Prod. 2012, 27, 11–20, doi:10.1016/j.jclepro.2011.12.035.
15. Vázquez-Rowe, I.; Villanueva-Rey, P.; Moreira, M.T.; Feijoo, G. Environmental analysis of Ribeiro wine from a timeline perspective: Harvest year matters when reporting environmental impacts. J. Environ. Manage. 2012, 98, 73–83, doi:10.1016/j.jenvman.2011.12.009.
16. WWF, Cork Screwed? Environmental and Economic Impacts of the Cork Stoppers Market; Report, WWF, Rome, Italy, 2006.
17. Corticeira Amorim, Evaluation of the Environmental Impacts of Cork Stoppers versus Aluminium and Plastic Closures; Final Report, Corticeira Amorim, Mozelos, Portugal, 2008.
18. Boudaoud, N.; Eveleigh, L.; Ruledge, D. Reconnaissance des Arômes et Nez Électronique. Ingénierie Analytique pour la Qualité des Aliments; Report; Institut National de Recherche en Agronomie (INRA): Versailles, France, 2003.
19. Marin, A.; Jorgensen, E.; Kennedy, J.; Ferrier, J. Effects of bottle closure type on consumer perceptions of wine quality. Am. J. Enol. Viticult. 2007, 58, 182–191.
20. Forum oenologie. Available online: http://www.oenologie.fr/ (accessed on 15 October 2012).
21. European Commission, Joint Research Center, Institute for Environment and Sustainability. International Reference Life Cycle Data System (ILCD) Handbook—Framework and Requirements for Life Cycle Impact Assessment Models and Indicators; Publications Office of the European Union: Luxembourg, Luxembourg, 2010.
22. Riboulet, J.-M. Goût de bouchon: Le point sur les origines et les recherches. Rev. Oenolog. 2003, 53, 41–43.
23. Moreau, M. Les moisissures des bouchons. Acad. Agric. Fr. C. R. 1978, 64, 842–849.

24. Davis, C.; Fleet, G.; Lee, T. The microflora of wine cork. Aust.Grape Wine 1981, 208, 42–44.
25. Chatonnet, P. Etude de l'Australian Wine Research Institute sur les caractéristiques physico-mécaniques de différents obturateurs et la modification de la composition d'un vin blanc après 18 mois de conservation. Rev. Oenolog. 2002, 29, 11–16.
26. Lumia, G.; Perre, C. Les fluides supercritiques, une innovation au service du liège, Partie 1/2. Rev. Oenolog. 2005, 32, 12–15.
27. Lumia, G.; Aracil, J.M. Les fluides supercritiques, une innovation au service du liège, Partie 2/2. Rev. Oenolog. 2006, 33, 13–16.
28. Descout, J. Evolution des connaissances sur le bouchage en liège des vins, les bouchons composites. Rev. Oenolog. 2008, 35, 28–31.
29. MIS Trend. Etude sur le Marché du Vin en Suisse: Notoriété, Habitudes de Consommation et d'Achat, Image; Report; MIS Trend: Lausanne, Switzerland, 2008. Available online: http://www.mistrend.ch/articles/La_viticulture_suisse.pdf/ (accessed on 15 October 2012).
30. Struijs, J. SimpleTreat 3.0: A Model to Predict the Distribution and Elimination of Chemicals by Sewage Treatment Plants; Report; National Institute of Public Health and the Environment: Bilthoven, The Netherlands, 1996.
31. VROM. Protocol 8136 Afvalwater, t.b.v NIR 2008 uitgave maart 2008 6B: CH4 en N2O uit Afvalwater; Directie Klimaatverandering en Industrie: The Hague, the Netherlands, 2008.
32. Daelman, M.; van Voorthuizen, E.; van Dongen, U.; Volcke, E.; van Loosdrecht, M. Methane emission during municipal wastewater treatment. Water Res. 2012, 46, 3657–3670, doi:10.1016/j.watres.2012.04.024.
33. IPCC. Contribution of Working Group I to the Fourth Assessment Report of the Intergovernmental Panel on Climate Change; Cambridge University Press: Cambridge, UK, 2007.
34. Humbert, S.; Margni, M.; Jolliet, O. IMPACT 2002+: Methodology Description, Draft for Version 2.1. unpublished work.

CHAPTER 6

INFLUENCE OF WINEMAKING PRACTICES ON THE CHARACTERISTICS OF WINERY WASTEWATER AND WATER USAGE OF WINERIES

A. CONRADIE, G. O. SIGGE, AND T. E. CLOETE

6.1 INTRODUCTION

6.1.1 STATISTICS OF THE WINE INDUSTRY

Wine production plays a big role in the agricultural industry around the world. In 2012, a volume of 252. 9 x 10^6 hL of wine was produced worldwide (OIV, 2013). The topproducing wine countries are Australia, Chile and the United States, followed by Argentina, France, Germany, Italy, Spain and South Africa (SA) (Devesa-Rey et al., 2011), with SA producing 10 x 10^6 hL of wine (OIV, 2013).

Table 1 shows the number of wineries in SA per production category based on the volume of grapes crushed, which ranges from five tons to 75 000 tons of grapes per harvest. The average winery crushes between one and 100 tons of grapes. White wine production makes up more than 70% of South African wine production (SAWIS, 2013).

Influence of Winemaking Practices on the Characteristics of Winery Wastewater and Water Usage of Wineries. © *Conradie A, Sigge GO, Cloete TE.* South African Journal of Enology and Viticulture, *35,1 (2014).*

6.1.2 COMPOSITION OF GRAPE JUICE AND WINE

The composition of grape juice and wine is compared in Table 2. There is almost no difference in the compounds found, other than their concentration, although additional compounds are formed during the winemaking process. Some of these compounds have to be removed before bottling. Fermentable sugars are transformed to alcohol according to the variety and the ripeness of the grapes; this is the most important difference between grape juice and wine (Stevenson, 2007).

TABLE 1: Number of wineries in South Africa per production categoryin 2012 (SAWIS, 2013)

Category	(tons of grapes crushed) Number of wineries
1 – 100	259
100 – 500	159
500 – 1 000	52
1 000 – 5 000	59
5 000 – 10 000	16
> 10 000	39

6.1.3 WINEMAKING PROCESS

The fundamentals of winemaking have stayed the same since biblical times (Hands & Hughes, 2001). What has changed is our ability to maintain the sterile environment required to produce top-quality wine (Halliday & Johnson, 1994). It is important to understand the winemaking process when looking into the quantities of wastewater produced at wineries. Figure 1 presents a schematic diagram of the major steps in winemaking and where waste is produced. The waste that is produced during the first step (destemming) is easily separated from the water, thus this is the only step that does not contribute directly to the chemical oxygen demand (COD) in the raw wastewater (Woodard & Curran, 2006).

TABLE 2: Composition of fresh grape juice and wine (Adapted from Stevenson, 2007).

Component	Grape juice (percentage by volume)	Wine (percentage by volume)
Water	73.5	86
Carbohydrates	25	0.2
Cellulose	5	-
Sugar	20	-
Alcohol (ethyl alcohol	-	12
Glycerol	-	1
Organic acids	93	35
Tartaric acid	0.54	0.20
Malic acid	0.25	-
Lactic acid	-	0.15
Citric acid (plus traces of succinic and lactic acid)	0.01	-
Succinic acid (plus traces of citric and malic acid)	-	0.05
Minerals	0.5	0.2
Calcium	0.025	0.02
Chloride	0.01	0.01
Magnesium	0.025	0.02
Potassium	0.25	0.075
Phosphate	0.05	0.05
Silicic acid	0.005	0.005
Sulphate	0.035	0.02
Others	0.1	Traces
Tannin and colour pigments	0.13	0.1
Nitrogenous matter	0.07	0.025
Amino acids	0.05	0.01
Protein and other nitrogenous matter	0.02	0.015
Volatile acids (mostly acetic acid	-	0.045
Esters	-	0.025
Aldehydes	-	0.004
Higher alcohols	-	0.001
Vitamins	Traces	Traces

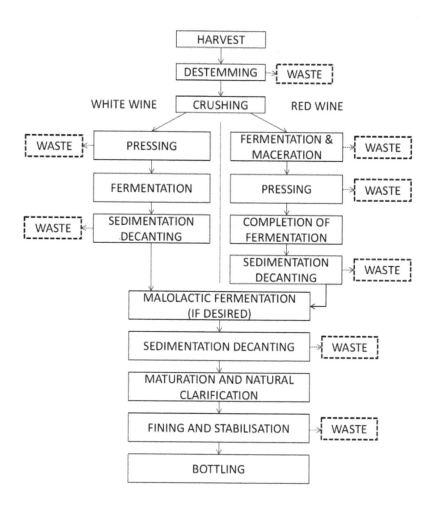

FIGURE 1: Diagram of organic waste generated in the process of making red and white wine (adapted from Arvanitoyannis et al., 2006; Devesa-Rey et al., 2011).

TABLE 3: General winemaking practices during the different wine-production periods in wineries for South Africa (adapted from South Australian EPA, 2004).

Period	Months	Action in cellar
Pre-harvest	Beginning to middle of January	Caustic washing of tanks and equipment, non-caustic washing of equipment in preparation for vintage
Early harvest	Middle to end of January	Wastewater production is rapidly increasing and has reached 40% of the maximum weekly flow. Vintage operations dominated by white wine production
Peak harvest	February and March	Wastewater generation is at its peak, vintage operations are at a maximum
Late harvest	Beginning of April	Wastewater production has decreased; vintage operations are dominated by red wine production
Post-harvest	End of April and May	Vintage operations have ceased. Caustic washing of the tanks and equipment used during the harvest
Non-harvest	June	Filtering of white wines in preparation for bottling. Filtering earth residues in waste water
Non-harvest	July	Cleaning bottling equipment with caustic. Bottling white wines
Non-harvest	August, September and October	Put red wine to barrel and filtering of previous year's reds. Water use is low
Non-harvest	November, December and beginning of January	Cleaning bottling equipment with caustic. Bottling wines

6.2 WATER USE IN A WINERY

Winemaking is seasonal and most of the activities related to it occur during the harvest period (Guglielmi et al., 2009). In the Southern Hemisphere, harvest is from the end of January to the beginning of April (Hands & Hughes, 2001). Throughout the year, the water volume and pollution load vary in relation to the different processes taking place (Arienzo et al., 2009a). Large volumes of polluted water are produced by winemaking and may vary from one winery to another, depending on the production period and the unique style of winemaking of the different wineries (Agustina et al., 2007). A big difference can be found when comparing the water use of different wineries due to parameters such as the type of tanks, processing equipment and various winemaking techniques (Walsdorff et al., 2004).

Table 3 describes the different periods and winemaking practices during the year that contribute to the volume and quality of winery wastewater. Generally, the pre-vintage period (beginning to middle of January) is used to clean the cellar and equipment in preparation for the harvest. This is essential to prevent the growth of micro-organisms on the equipment, which can lead to contamination of the juice (Mercado et al., 2006). Due to the regular/daily cleaning of equipment during the harvesting period (end of January to beginning of April), there is a bigger demand for clean water (Rodriguez et al., 2007). After harvesting, hygiene is still an immense priority, despite the decrease in the volume of clean water used (due to activities in the cellar.) During the postharvest period, it is possible that there may be days without water usage in the wine cellar (P. Ngamane, Assistant winemaker, Hartenberg Wine Estate, Stellenbosch, personal communication, 2012).

TABLE 4: Estimates of volumes of virtual water used to produce wine.

Volume of water per litre of wine produced	Estimated volume of total water used for the wine industry worldwide	Reference
5 – 8	$1.3 – 2.1 \times 10^9$ hL	Mosse et al. (2011)
1 – 4	$2.6 – 10.5 \times 10^7$ hL	Bolzonella et al. (2010)
0.97 – 1.25	$2.5 – 3.3 \times 10^7$ hL	Lucas et al. (2010)

In the winter months (rainy season) it is important that the storm water and winery wastewater are separated to prevent an increase in the amount of water that needs to be treated. It is also vital that the storm water remains unpolluted (Walsdorff et al., 2004).

The term 'virtual water' was first used by Allan in 1997 to describe water embedded in water-intense commodities (Allen, 1997; Wichelns, 2001). Since then the term has been widely used to describe the volume of fresh water used to produce a product, in this case wine, from the beginning of the process, through harvest, right to the end where it is in the bottle and ready for trade and consumption. Virtual water is being used to calculate the impact of the water footprint of specific production systems (Herath et al., 2012).

The virtual water used to produce one litre of wine varies according to different literature sources from around the world. Table 4 provides a summary of estimates of global winery water-use volumes according to the Organisation Internationale de la Vigne et du Vin (OIV, 2011) of wine produced in 2010. It is clear that there is a significant difference between the respective estimates.

Furthermore, the wine industry in South Africa has grown by 58% since 1997, from 5.5 x 10^6 hL in 1997 to 8.7 x 10^6 hL in 2012 (Fig. 2). This is a significant increase in wine, and goes hand in hand with the volume of water used and consequently the wastewater generated for every litre of wine produced (SAWIS, 2013).

6.3 COMPOSITION OF WINERY WASTEWATER

One of the biggest issues for the wine industry is the management of large volumes of wastewater (Bustamante et al., 2005). While wine production does not have a reputation as a polluting industry, the wastewater volumes worldwide are increasing and the wastewater has a high organic load, low pH, variable salinity and nutrient levels, all of which indicate that the wastewater has the potential to pose an environmental threat (Mosse et al., 2011). The four biggest components contributing to wastewater pollution in a winery are:

1. Sub-product residues: stems, skins, sludge, lees, tartar (Musee et al., 2005).
2. Lost brut production: must and wine occurred by spillage during winemaking activities (Mosse et al., 2011).
3. Products used for wine treatment: fining agents and filtration earths (Pérez-Serradilla & Luque de Castro, 2008).
4. Cleaning and disinfection products (e.g. sodium hydroxide (NaOH) and potassium hydroxide (KOH) used to wash materials and equipment (Mahajan et al., 2010).

Table 5 shows the influence on wastewater of the different steps in the winemaking process.

An analysis of the average characteristics of wastewater showed that winery wastewater around the world and in different wineries in the same country has significant differences (Mosse et al., 2011). A summary of data for a few wineries is given in Table 6 to illustrate the differences in wastewater characteristics in different studies. The variance in wastewater composition complicates the issue of finding a general solution for wastewater treatment at different wineries (Andreottola et al., 2009). To find the correct treatment and reuse efficiencies for wastewater it is important to understand the detailed composition of the wastewater (Bustamante et al., 2005).

TABLE 5: Winery actions related to winery wastewater quantity and quality and the impact on the quality parameters (adapted from Van Schoor, 2005

Winery action	Impact on wastewater quantity	Impact on wastewater quality	Impact on legal wastewater quality parameters
Cleaning water			
Alkali washing and neutralisation	Up to 33%	Increase in NA, K, COD and pH	Increase in EC, SAR, COD, variation in pH
Rinse water (tanks, floors, transfer lines, bottles, barrels, etc.)	Up to 43%	Increase in NA, P, Cl, COD	Increase in EC, SAR, COD, variation in pH
Process water			
Filtration with filter aid	Up to 15%	Various contaminants	Increase in COD and EC
Acidification and stabilisation of wine	Up to 3%	H2SO4 or NaCl	Increase in COD and EC, decrease in pH
Cooling tower waste	Up to 6%	Various salts	Increase COD and EC
Other sources			
Laboratory practices	Up to 5–10%	Various salts, variation in pH, etc.	Increase COD and EC

EC – electrical conductivity; SAR – sodium absorption rate; COD – chemical oxygen demand

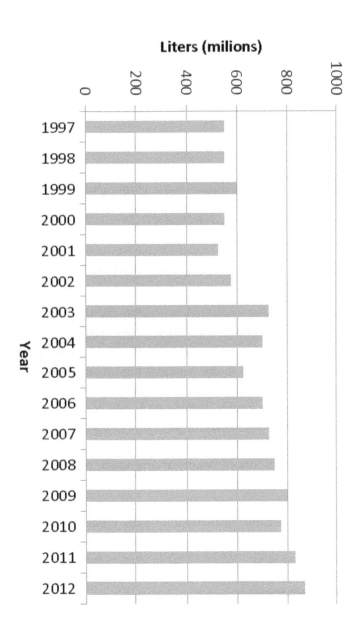

FIGURE 2: Wine production volumes in South Africa from 1997 to 2012 (SAWIS, 2013).

TABLE 6: Summary of reported winery wastewater characteristics.

Parameters	Unit	Min	Max	Mean	References
COD	mg/L	340	49105	14426	1–10
BOD	mg/L	181	22418	9574	4, 6, 7, 10
pH	-	3.5	7.9	4.9	2, 4, 6, 8, 9, 10
Total solids	mg/L	190	18000	4151	2, 4, 5, 8
EC	S/m	1.2	7.2	4.16	2, 4, 6, 8
Suspended solids	mg/L	1000	5137	2845	4, 9, 10

The reference numbers in the last column refer to the following: 1. Agustina et al. (2007); 2. Arienzo et al. (2009b); 3. Bolzonella et al. (2010); 4. Bustamante et al. (2005); 5. Eusebi et al. (2009); 6. Mahajan et al. (2010); 7. Rodriguez et al. (2007); 8. Rytwo et al. (2011); 9. Yang et al. (2011); 10. Zhang et al. (2006)

6.3.1 ORGANIC COMPOUNDS IN WINERY WASTEWATER

Most of the wastes generated in a cellar (80 to 85%) are organic wastes (Ruggieri et al., 2009). The organic material in winery wastewater is generated from the grapes and wine (Valderrama et al., 2012). Figure 1 illustrates the points in the winemaking process where organic material contributes to the composition of winery wastewater. After destemming and pressing the grapes, (white and red) grape marc is produced that consists of grape skins and pips (Devesa-Rey et al., 2011). Despite the fact that the skins are kept separate from the wastewater system, the residue on the floors of the cellar and in the press will contribute to the high levels of COD (chemical oxygen demand) and variation in pH (Van Schoor, 2005). Apart from this, lees will form on the bottom of the wine tank or barrels after fermentation of the grape juice. This sediment will also have an effect on the organic compounds and COD of the wastewater (Mosse et al., 2011). COD is used to measure the oxygen demand of the organic load present in the wastewater (Andreottola et al., 2009). COD levels for grape marc can range from 15 000 to 44 900 mg/L, for lees from 27 200 to 36 100 mg/L and for wine from 26 200 mg/L (Fillaudeau et al., 2008).

Bories et al. (1998, cited in Fillaudeau et al., 2008) studied the composition and components portions of winery wastewater. Their study showed that 90% of the organic component is ethanol, except during harvest, when it is mainly sugars (Fillaudeau et al., 2008). Ethanol concentrations of 4 900 mg/L and sugar (glucose and fructose) of 870 mg/L were detected in winery wastewater with a dissolved COD of 12 700 mg/L (Table 7). In addition to this, a study by Colin et al. (2005) showed that there is a linear correlation between COD and the ethanol concentration, and therefore the organic load of winery wastewater can be estimated when the ethanol concentration is known.

Contributing to the difference in the composition of the organic material in wastewater are the uncontrolled chemical reactions that take place in the wastewater (Mosse et al., 2011). Organic acids (acetic, tartaric, malic, lactic and propionic), alcohols, esters and polyphenols play an important role in the composition of winery wastewater (Zhang et al., 2006; Mosse et al., 2012).

There is not a lot of research available on the organic components of winery wastewater, but it is essential to characterise the organic composition of winery wastewater to establish the impacts the wastewater will have on the environment (Bustamante et al., 2005; Mosse et al., 2011).

6.3.2 INORGANIC COMPOUNDS IN WINERY WASTEWATER

The composition of the inorganic compounds in winery wastewaters is dependent mainly (up to 76%) on the components of the cleaning agents used in wineries (Table 5), except for potassium, which is present in high concentrations in grape juice (Mosse et al., 2011). Strong alkaline-based cleaning agents that are good for tartrate removal include caustic soda (NaOH) and caustic potash (KOH) (Sipowicz, 2007). Wineries that uses sodium-based cleaning agents have problems with the salinity of the wastewater if it is used for irrigation. The inorganic ions present are predominantly potassium and sodium, with low levels of calcium and magnesium, although the concentrations of both organic and inorganic constituents vary with differences in winemaking operations over time, as well as between individual wineries (Mosse et al., 2012).

6.4 WHY MANAGE WASTE/WASTEWATER?

In the past, the small volumes of winery wastewater that were produced by wineries had little effect on the immediate environment, but with increasing wine production all around the world, winery wastewater is a rising concern for the contamination of subsurface flow, soil and the environment (Grismer et al., 2003).

Research on the composition and volumes of winery wastewater is receiving more attention, and awareness of the effects of winery wastewater is assisting in the establishment and improvement of winery wastewater treatment systems (Devesa-Rey et al., 2011). Moderate quantities of winery waste and wastewater that are exposed to soils can increase the organic material due to the high concentration of soluble organic carbon in winery wastewater, which, in turn, will enhance the fertility of the soils (Bustamante et al., 2011). Unfortunately, continuous exposure to the organic material can lead to organic overload that blocks the pores and lowers the quality of the soils immensely (Vries, 1972). The continuous addition of winery wastewater to soils can also contribute to high soil salinity, which can lead to dispersion (Halliwell et al., 2001).

The disposal of grape marc, a complex lignocellulose material made up of the skin, stalks and seeds, has also been a problem for wineries. In total, more than 20% of wine production is waste, comprising thousands of tons (Arvanitoyannis et al., 2006). Untreated grape marc can lead to several environmental threats, including foul odours and ground water pollution (Table 8). Decomposing grape marc provides the perfect environment for flies and pest to flourish (Laos et al., 2004). Leachate from the marc contains tannins and other chemical compounds that infiltrate the surface soil and ground water, leading to oxygen depletion (Arvanitoyannis et al., 2006). It is possible to use the marc in other industries (Kammerer et al., 2005); however, this can be expensive and therefore other, alternative solutions must be found (Ruggieri et al., 2009). The impact of winery wastewater on the biological and physiochemical properties of soil has not been researched in depth (Mosse et al., 2012). Table 8 shows the potential impacts of winery wastewater on the environment.

6.5 MINIMISATION OF WATER USAGE AND POLLUTION LOAD

Before discussing the different treatment options it is important to understand that the minimisation of winery wastewater should be the goal of all wineries (Lee & Okos, 2011). The term 'zero discharge process' is used by Lee and Okos (2011) to refer to the substantial reduction of water and energy usage and ultimately to generate no waste during the production of food and beverages. Furthermore, water saving does not only cut the cost of fresh water used, but also reduces the cost accompanying the treatment of the wastewater (Fillaudeau et al., 2008). Avoiding waste is the most costeffective and often the easiest principle to implement – better known as 'prevention (waste minimisation/cleaner production) is better than cure' (Chapman et al., 2001).

TABLE 7: Composition and breakdown of the COD of winery wastewater (Adapted from Fillaudeau et al., 2008)

	Concentration (mg/L)	COD (%)
COD raw	14 600	
COD dissolved	12 700	100
Ethanol	4 900	80.3
Glucose and fructose	870	7.3
Glycerol	320	3.1
Tartaric acid	1 260	5.3
Malic acid	70	0.4
Lactic acid	160	1.2
Acetic acid	300	2.6

Not only is water a limited resource, but it can also contribute to the total cost of the final product. When the total cost of production water is calculated for the food and beverage industry it is vital not just to look at the cost of the volume used and the volume disposed, but also to look at the potential loss in income when the product is disposed as effluent (Casani et al., 2005). Fillaudeau formulated it as follows:

Water cost savings = $W_{Saved} \times (W_{Rate} + W_{Treat})$ where W_{Saved} is the volume of water saved, W_{Rate} the incoming water rate (e.g. $/m^3$), and W_{Treat} the treatment cost of the incoming water (Fillaudeau et al., 2008).

Wastewater treatment charge = $WW_{Saved} \times WW_{Charge}$ where WW_{Saved} is the volume of wastewater saved and WW_{charge} is the volumetric wastewater charge, which is generally a function of several parameters such as COD, total solid content as well as other specific components (Fillaudeau et al., 2008).

In Figure 3, the principles of cleaner production are illustrated, with the most preferred option, avoidance, as the most important principle (Chapman et al., 2001).

Water management is a particular concern in the wine industry and there are practices that can be implemented to help reduce the wastewater volumes of wineries by using cleaner production principles (Van Schoor, 2005). In general, a considerable volume of up to 30% can be reduced through simple changes with minimum capital input (Kirby et al., 2003). These changes include the evaluation of water usage in controlled areas;

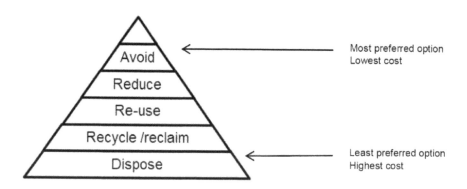

FIGURE 3: Hierarchy of cleaner production principles (Chapman et al., 2001).

the improvement of planning and the control of water use; the option to reuse water; water recycling after treatment; and, lastly, the improvement of the layout of the processing area (Klemeš et al., 2009). In particular, the evaluation (water auditing) of water usage is important to all industries (Klemeš et al., 2008). Water auditing will not only help the winery to understand where the water is used, but also will indicate the place/process of largest usage. More importantly, it will point out the areas of unnecessary waste (Klemeš et al., 2008).

TABLE 8: Potential environmental impacts of winery wastewater (adapted from South Australian EPA, 2004).

Winery wastewater components	Indicators	Effects
Organic matter	BOD, TOC, COD	Reduces oxygen levels – death of fish and other aquatic organisms. Odours generated by anaerobic decomposition
Alkalinity/ acidity	pH	Death of aquatic organisms at extreme pH. Affects the solubility of heavy metals in the soil and availability and/or toxicity in water affects crop growth
Nutrients	N, P, K	Eutrophication or algal bloom. N as nitrate and nitrite in drinking water supply can be toxic to infants
Salinity	EC, TDS	Imparts undesirable taste to water, toxic to aquatic organisms, affects water uptake by crops
Sodicity	SAR	Affects soil structure, resulting in surface crusting. Low infiltration and hydraulic conductivity
Heavy metals	Cu, Ni, Pb, Zn, Hg, etc.	Toxic to plants and animals
Solids	TSS	Can reduce light transmission in water, thus compromising ecosystem health, smothers habitats, odour generated from anaerobic decomposition

BOD – biochemical oxygen demand; TOC – total organic carbon; TDS – total dissolved solids; TSS – total suspended solids

    The first step for all wineries should be to install a water meter to control their water usage and identify water usage peaks, and to seek possible solutions (water-saving practices) to decrease water usage (Chapman,

1996). These water-saving practices include the use of nozzles on water pipes to avoid wastage. Water can be turned on and off conveniently each time the handle is compressed. The dry sweeping of floors with brushes and squeegees will also ensure the reduction of water usage before the floors are washed (Walsdorff et al., 2004). Furthermore, water-saving practices that could be introduced to winery staff with the appropriate training are listed in Table 9. In addition to these principles, it is vital that the management is 100% committed. Winewatch also recommends that all winery staff should be involved when a cleaner production strategy is developed, as this will heighten their awareness (Klemeš et al., 2008; Anonymous, 2009).

TABLE 9: Water-saving practices (Chapman, 1996; Walsdorff et al., 2004).

Water-saving practices	Description
Installation of water meter	Control water usage and identify water usage peaks
Use minimum water	Use no more water than needed for the job
High-pressure water system	Less water required for more efficient cleaning
Nozzle on water pipes	Avoid wastage of water so that a hose will not run when not required
Use of brushes and squeegee	Dry sweeping of floors before washing
Water awareness training	Developing of a cleaner production strategy

The next step in minimising the pollution load of the winery wastewater is the reduction of organic material in the raw wastewater. A number of practices that wineries can implement to achieve this is listed in Table 10. Firstly, the installation of mesh sieves in the floor, especially in the grape-processing section, will prevent organic material (stalks, skins and pips) from entering the wastewater system. Secondly, installing in-line screening in the initial stages of the treated wastewater system will reduce finer solids in the wastewater. To add to this, the accumulated lees after the initial alcoholic fermentation is thick and spillage can easily increase the COD (Fillaudeau et al., 2008) of the generated wastewater. The average COD of wine lees is 76 000 mg/L (Boires et al.,1998, cited in Fillaudeau et al., 2008), therefore a small volume (of spillage when the tank is washed out after use) could have a vast influence on the raw wastewater of the winery. Thus it would be better if the lees and the first rinsing of the lees are transferred to a separate tank to prevent the lees and diluted lees from draining into the wastewater system (Woodard & Curran, 2006). It goes without saying that lees that is more compactly settled will simplify this last-mentioned practice. The use of fining agents that produce more compactly settled lees should be considered by the winemaker to help reduce the volume of the lees (Anonymous, 2009). Chapman (1996) has drawn attention to the fact that avoidance is an important principle to consider when planning to reduce wastewater pollution, therefore, transfers of wine should be kept to a minimum to reduce the chances of spillage.

TABLE 10: Pollution load-minimisation practices (Chapman et al., 1996; Woodard & Curran, 2006; Winewatch, 2009).

Pollution load-minimisation practice	Description
Installing mesh sieves	Prevent organic matter in winery wastewater
Pomace animal feed/fertiliser	Mixed with stems and other solids
Transfer lees and first rinse to separate tank	Prevent the lees and diluted lees from draining into the wastewater system
Ensure that conveyers, storage bins and tanks are not overfilled	Reduce spillage
Grape seed oil	Edible oils can be extracted form grape seeds
Use fining agents that produce most compact lees	Reduce volume of lees
Install in-line screening of organic matter	Reduce finer solids in wastewater
Recovery of tartrates	Use in cooking as cream of tartar
Resettle lees	Remove as much organic material as possible
Keep transfers to a minimum	Reduce changes of spillage

Primarily, the elimination of salt (K, Ca, Na & Mg) usage in the winery should be promoted to reduce the EC and sodium absorption rate (SAR); consequently no treatment would be necessary before irrigation with the wastewater. The use of non-sodium-based cleaning chemicals is advised by Chapman (Chapman, 1996). Replacing disinfectants and cleaning agents with ozone will result in lowering the EC and COD (Van Schoor, 2005). The initial cleaning with caustic can also be substituted with a high-pressure rinse or with heat/steam (Anonymous, 2009). When caustic is used for cleaning, the aim should be to re-use it (Chapman, 1996).

6.6 CONCLUSIONS

Wine production is a growing industry all around the world because the demand for wine is increasing and new wineries are being established. Subsequently, this industry produces large volumes of wastewater that potentially pose a risk to the environment and thus require treatment.

It is clear from the literature that the volumes of water used in wineries vary, but also that the wastewater characteristics differ significantly. This is an indication that the winemaking practices (white, red, rosé or blends; type of press; bottling operations; filtering and barrel work; to name a few) influence the wastewater characteristics. Contributing to this problem is that many winemakers see winemaking as an art and thus are not overly concerned with water use and wastewater treatment.

Even though the characteristics of wastewater produced by wineries differ immensely from winery to winery, there are practices that wineries can implement to reduce the volume and the pollution load. The implementation of cleaner production practices offers a partial solution for wineries to minimise wastewater produced and also reduce their water usage. It goes without saying that this, in return, will potentially reduce cost by avoiding wastages. Apart from these principles it is vital that the management of the winery is absolutely committed to contributing to the awareness and motivation of their employees.

A number of studies have recently shown that there is a correlation between winemaking practices and the characteristics of the wastewater, but further investigation is required to elucidate how specific winery practices influence the characteristics/composition of winery wastewater.

More specific information on these practices and their effects might encourage wineries to implement more efficient practices, thereby reducing water usage and the pollution loads of winery wastewater.

REFERENCES

1. Agustina, T.E., Ang, H.M. & Pareek, V.K., 2007. Treatment of winery wastewater using a photocatalytic/photolytic reactor. Chem. Engin. J. 135, 151-156.

2. Andreottola,AD G., Foladori, P. & Ziglio, G., 2009. Biological treatment of winery wastewater: An overview. Water Sci. Technol. 60(5), 1117-1125.

3. Anonymous, 2009. Winewatch fact sheet 2. [Internet document]. Available http://environmentagriculture.curtin.edu.au/local/docs/winewatch/Winewatch_Fact_Sheet_2.pdf (accessed 09/01/2012).

4. Arienzo, M., Christen, E.W. & Quayle, W.C., 2009a. Phytotoxicity testing of winery wastewater for constructed wetland treatment. J. Hazardous Mat. 169, 94-99.

5. Arienzo, M., Christen, E.W., Quayle, W. & Di Stefano, N., 2009b. Development of a low-cost waste water system for small-scale wineries. Water Environ. Res. 81(3), 233-242.

6. Arvanitoyannis, I.S., Ladas, D. & Mavromatis, A., 2006. Review: Wine waste treatment methodology. Int. J. Food Sci. Technol. 41, 1117-1151.

7. Bolzonella, D., Fatone, F., Pavan, P. & Cecchi, F., 2010. Application of a membrane bioreactor for winery wastewater treatment. Water Sci. Technol. 62(12), 2745-2759.

8. Bustamante, M.A., Paredes, C., Moral, R., Moreno-Caselles, J., Perez-Espinosa, A. & Parez-Murcia, M.D., 2005. Uses of winery and distillery effluents in agriculture: Characterisation of nutrient and hazardous components. Water Sci. Technol. 51(1), 145-151.

9. Bustamante, M.A., Moral, R., Paredes, C., Peres-Espinosa, A., Moreno-Caselles, J. & Perez- Murcia, M.D., 2007. Agrochemical characterisation of the solid products and residues from the winery and distillery industry. Waste Management 28(2), 372-380.

10. Bustamante, M.A., Said-Pullicino, D., Agulló, E., Andreu, J., Paredes, C. & Moral, R., 2011. Application of winery and distillery waste composts to a Jumilla (SE Spain) vineyard: Effects on the characteristics of a calcareous sandy-loam soil. Agric. Ecosys. Environ. 140, 80-87.

11. Casani, S., Rouhany, M. & Knøchel, S., 2005. A discussion paper on challenges and limitations to water reuse and hygiene in the food industry. Water Res. 39, 1134-1146.

12. Chapman, J.A., 1996. Cleaner production for the wine industry. South. Australian Wine and Brandy Industry Association, Adelaide, Australia, pp 1-31.

13. Chapman, J.A., Baker, P. & Willis, S., 2001. Winery wastewater handbook: Production, impacts and management. Winetiles, Adelaide.

14. Colin, T., Boires, A., Sire, Y. & Perrin, R., 2005. Treatment and valorisation of winery wastewater by a new biophysical process (ECCF). Water Sci. Technol. 51(1), 99-106.

15. Devesa-Rey, R., Vecino, X., Varela-Alenda, J.L., Barral, M.T., Cruz, J.M. & Moldes, A.B., 2011. Valorization of winery waste vs cost of not recycling. Waste Management 31, 2327-2335.

16. Eusebi, A.L., Nardelli, P., Gatti, G., Battistoni, P. & Cecchi, F., 2009. From conventional activated sledge to alternate oxic/anoxic process: the optimisation of winery wastewater treatment. Water Sci. Technol. 60(4), 1041-1048.

17. Fillaudeau, L., Bories, A. & Decloux M., 2008. Brewing, winemaking and distilling: An overview of wastewater treatment and utilisation schemes. In: Handbook of water and energy management in food processing, Volume 1. Woodhead Publishing, pp 83-112.

18. Grismer, M.E., Carr, M.A. & Shepherd, H.L., 2003. Evaluation of constructed wetland treatment performance for winery wastewater. Water Environ. Res. 75(5), 412-421.

19. Guglielmi, G., Andreottola, G., Foladori, P. & Ziglio, G., 2009. Membrane bioreactors for winery wastewater treatment: Case studies at full scale. Water Sci. Technol. 60(5), 1201-1206.

20. Halliday, J. & Johnson, H., 1994. Making white and red wine. In: The art and science of Wine. Mitchell Beazley, London, pp 88 – 142. Halliwell, D., Barlow, K. & Nash, D., 2001. A review of the effects of wastewater sodium on soil physical properties and their implications for irrigation systems. Soil Res. 39, 1259-1267.

21. Hands, P. & Hughes, D., 2001 (2nd ed). How wine is made. In: New world of wine from the Cape of Good Hope. The definitive to the South African wine industry. Stephen Phillips, Somerset West, pp 84 – 91.

22. Herath, I., Green, S., Singh, R., Horne, D., Van der Zijpp, S. & Clothier, B., 2012. Water footprinting of agricultural products: A hydrological assessment for the water footprint of New Zealand's wines. J. Cleaner Production. doi:10.1016/j. jclepro.2012.10.024

23. Kammerer, D., Kljusuric, J.G., Carle, R. & Schieber, A., 2005. Recovery of anthocyanins from grape pomace extracts (Vitis vinifera L. cv. Cabernet Mitos) using a polymeric adsorber resin. Eur. Food Res. Technol. 220, 431-437.

24. Kirby, R.M., Bartram, J. & Carr, R., 2003. Water in food production and processing: Quantity and quality concerns. Food Control 14, 283-299.

25. Laos, F., Semenas, L. & Labud, V., 2004. Factors related to the attraction of flies at a biosolids composting facility (Bariloche, Argentina). Sci Tot. Environ. 328, 33-40.

26. Klemeš, J., Smith, R. & Kim, J. (2008). Assessing water and energy consumption and designing strategies for their reduction. In: Handbook of Water and Energy Management in Food Processing. Woodhead, pp 83-105.

27. Klemeš, J.J., Varbanov, P.S. & Lam, H.L. (2009). Water footprint, water recycling and food industry supply chains. In: Handbook of Waste Management and Co-Product Recovery in Food Processing. Woodhead, pp 134-168.

28. Lee, W.H. & Okos, M.R., 2011. Sustainable food processing systems – Path to a zero discharge: Reduction of water, waste and energy. Food Sci. 1, 1768-1777.

29. Lucas, M.S., Peres, J.A. & Puma, G.L., 2010. Treatment of winery wastewater by ozone-bases advanced oxidation processes (O3, O3/UV and O3/UV/H2O2) in a pilot-scale bubble column reactor and process economics. Separation & Purification Tech. 72, 235-241.

30. Mahajan, C.S., Narkhede, S.D., Khatik, V.A., Jadhav, R.N. & Attarde, S.B., 2010. A review: Wastewater treatment at winery industry. Asian J. Environ. Sci. 4(2), 258-265.

31. Mercado, L., Dalcero, A., Masuelli, R. & Combina, M., 2006. Diversity of Saccharomyces strains on grapes and winery surfaces: Analysis of their contribution to fermentative flora of Malbec wine from Mendoza (Argentina) during two consecutive years. Diversity Food Microbiol. 24, 403-412.

32. Mosse, K.P.M., Patti, A.F., Christen, E.W. & Cavagnaro, T.R., 2011. Review: Winery wastewater quality and treatment options in Australia. Aus. J. Grape Wine Res. 17(2), 111-121.

33. Mosse, K.P.M., Patti, A.F., Smernik, R.J., Christen, E.W. & Cavagnaro, T.R., 2012. Physicochemical and microbiological effects of long- and shortterm winery wastewater application to soils. J. Hazardous Mat. 201-202, 219-228.

34. Musee, N., Lorenzen, L. & Aldrich, C., 2005. Cellar waste minimization in the wine industry: A systems approach. J. Cleaner Production 15, 417-431. OIV (Organisation Internationale de la Vigne et du Vin), 2013. Available http:// www.oiv.int/oiv/info/enpublicationsstatistiques (accessed 24/06/2013).

35. Pérez-Serradilla, J.A. & Luque de Castro, M.D., 2008. Role of lees in wine production: A review. Food Chem. 111, 447-456.

36. Rodriguez, L., Villasenor, J., Buendia, I.M. & Fernandez, F.J., 2007. Re-use of winery waste waters for biological nutrient removal. Water Sci. Technol. 56(2), 95-102.

37. Ruggieri, L., Cadena, E., Martínez-Blanco, J., Gasol, C.M., Rieradevall, J., Gabarrell, X., Gea, T., Sort, X. & Sánchez, A., 2009. Recovery of organic wastes in the Spanish wine industry. Technical, economic and environmental analyses of the composting process. J. Cleaner Production 17, 830-838.

38. Rytwo, G., Rettig, A. & Gonen, Y. (2011). Organo-sepiolite particles for the efficient pretreatment of organic wastewater: Application to winery effluents. Applied clay Sci. 51, 390-394.

39. SAWIS (South African Wine Industry and Systems), 2013. Available http://www.sawis.co.za/info/download/Book_2013_eng.pdf (accessed 24/06/2013).

40. Sipowicz, M., 2007. Winery cleaning and sanitation. Texas cooperative extension. Available http://winegrapes.tamu.edu/winemaking/ Sanitation%20Guide.pdf (accessed 21/12/2012).

41. South Australian Environment Protection Authority (EPA), 2004. Guidelines for Wineries and Distilleries. [Internet Document] Available http://www.epa.sa.gov.au/xstd_files/Industry/Guideline/guide_wineries.pdf (accessed 14/06/2012).

42. Stevenson, T., 2007. How wine is made. In: The new Sotheby's wine encyclopaedia: A comprehensive reference guide to the wines in the word. Dorling Kindersley Limited, London, pp 32 – 38.

43. Van Schoor, L.H., 2004. A prototype ISO 14001 Environmental Management System for wine cellars. PhD dissertation, Stellenbosch University, Private Bag X1, 7602 Matieland (Stellenbosch), South Africa.

44. Van Schoor, L.H. (2005). Winetech: Guidelines for the management ofwastewater and solid waste atexisting wineriesURL: http://www.ipw.co.za/content/guidelines/WastewaterApril05English.pdf. (12/04/2012)

45. Vries, J.D., 1972. Soil filtration of wastewater effluent and the mechanism of pore clogging. J. Water Poll. 44, 565-573.

46. Walsdorff, A., Van Kraayenburg, M. & Barnardt, C.A., 2004. A multisite approach towards integrating environmental management in the wine production industry. Water SA. 30(5), 82-87.

47. Wichelns, D., 2001. The role of 'virtual water' in efforts to achieve food security and other national goals, with an example from Egypt. Agric. Water Managem. 49, 131-157.

48. Woodard & Curran, 2006 (2nd ed). The winemaking industry. In: Industrial waste treatment handbook. Elsevier Inc, Oxford, pp 455 – 459.

49. Yang, R., Ma, Y., Zhang, W., Xu, R., Yin, F., Li, J., Chen, Y., Liu, S. & Xu Y., 2011. The performance of new anaerobic filter process for high concentration winery wastewater treatment. 978-4244-6255-1/11. Power and Energy Engineering Conference (APPEEC), 2011 Asia-Pacific.

50. Zhang, Z.Y., Jin, B., Bai, Z.H. & Wang, X.Y., 2006. Production of fungal biomass protein using microfungi from winery wastewater treatment. Bioresource Technol. 99, 3871-3876.

CHAPTER 7

ECO-PREMIUM OR ECO-PENALTY? ECO-LABELS AND QUALITY IN THE ORGANIC WINE MARKET

MAGALI DELMAS AND NEIL LESSEM

7.1 INTRODUCTION

Eco-labels are part of a new wave of environmental policies that empha-size information disclosure as a tool to induce environmentally friendly behaviors by both firms and consumers (Dietz & Stern, 2002). The goal of eco-labels is to reduce the information asymmetry between producers and consumers over the environmental attributes of a good (Crespi & Marette, 2005; Leire & Thidell, 2005). Prominent examples of eco-labels include the USDA organic label for agricultural products, the Energy Star label for energy appliances, and the Forest Sustainable Stewardship label for lumber. The number of eco-labels programs on the market has proliferated from a mere dozen worldwide in the 1990s to more than 377 programs in 2010. The corresponding market for eco-labeled products has grown sig-nificantly in value over the same time period, with products like organic fruit and vegetables capturing 12 percent of the US market in 2010 (Organic Trade Association, 2011).

Eco-labels are often developed by government agencies and non-governmental organizations which are separate to the industries that produce and sell the eco-product. This third-party certification lends credibility to the eco-labels (D'Souza et al, 2006; Leire & Thidell, 2005; Nillson et al, 2004), but may result in eco-labels that do not meet the needs of, or are even detrimental to producers (de Boer, 2003; Rex and Bauman, 2007; Stern, 1999). This difference stems from the informational goals of producers and labelers. Producers wish to use information over environmental attributes to match their products to the needs of consumers (Peattie, 2001), whereas the third-parties who actually issue the labels aim to close the information asymmetry between producers and consumers (de Boer, 2003; Rex and Bauman, 2007; Stern, 1999). While these two goals may sometimes align, creating increased demand for eco-labeled products (Teisl et al, 2002; Bjorner et al, 2004), this is not always the case. Many studies have found that eco-labels do not deliver the desired message, resulting in consumers who are unsure of the extra value that the eco-label presents (Nillson et al, 2004; Yiridoe et al, 2005); who are confused by distinctions between different eco-labels (Leire & Thidell, 2005; Bhaskaran et al, 2006); who do not match the eco-label to environmental problems (Van Amstel, 2008; Teisl et al, 2004) and who associate the eco-label with negative product attributes (Delmas & Grant, 2010; Rivera, 2002).

In this paper we run a discrete choice experiment over eco-labeled and non-eco-labeled wine to investigate circumstances where eco-labels may send insufficient or undesired information to consumers. The US wine market is particularly suited for this type of investigation due to both institutional and product characteristics. Institutionally, the government agency responsible for food-related eco-labels, the USDA, has created two very similar eco-labels, one of which is legitimately associated with product concerns and one which is not. In terms of product, wine is a processed good, making the consumer health aspect of eco-labeling less clear and potentially obfuscating the link between consumer needs and environmental attributes. Moreover wine is a differentiated product with a variety of characteristics, all of which may interact with or cancel out the signal that the eco-label sends. In our study, 830 participants from across the United States made a series of choices, where they selected between hypothetically purchasing one of four graphical representations of wine bottles, or nothing. This method allowed us to randomly vary wine attributes, price and eco-label, thereby revealing the full range of consumers' preferences, rather than the subset circumscribed by existing market choices. This discrete-choice exercise was combined with a survey that allowed us to link attitudes, demographics and behavior to wine choices.

We find that consumers prefer eco-labeled wine at low prices, but prefer non-eco-labeled wine at high prices. Since price acts as a signal of quality in the wine industry (Lockshin et al, 2006; Mtimet & Albisu, 2006), we interpret this as meaning that consumers interpret eco-labels as a signal of low quality. Support for this is found when we separately examine the two different eco-labels, one of which is legitimately associated with quality concerns. We find that consumers who are unaware of the difference between the two labels penalize high priced wines bearing either eco-label. However, consumers who know the difference between the two eco-labels, only penalize high priced wines bearing the eco-label associated with quality concerns.

This paper contributes to the growing literature on information disclosure as an environmental policy tool, by showing the adverse effects of considering information alone a sufficient tool. Additionally, our findings present a valuable lesson for the policy makers who utilize and frame information disclosure policies. An eco-label premium is essential for an eco-industry to sustainably exist. Thus any eco-labeling initiative needs to ensure that it will deliver such premiums. Focusing purely on information asymmetries will not necessarily create eco-labels that align eco-products with the needs of consumers. Instead government organizations need to work with producers and marketers to ensure that eco-labels provide information that clearly communicate their value to consumers.

The remainder of the paper proceeds as follows. In section 2 we explore the existing literature on eco-labels. In Section 3 we present a model of eco-label choice that includes quality signals. Section 4 presents our testable hypotheses, while section 5 discusses the methodology behind our discrete choice exercise. In section 6 we develop an econometric framework for analyzing the experiment, the implementation of which is presented in section 7. Our results are shown in section 8 with concluding comments and discussion given in section 9.

7.2 INFORMATION POLICIES

Information disclosure policies are increasingly gaining prominence as a "new tool" in environmental management policies (Dietz & Stern, 2002). These policies augment or replace government regulation by publicly providing information that will presumably assist more cost effective private and legal forces (Delmas et al, 2010). Environmental information disclosure policies can be instituted at either the firm, product or consumer level. Firm level information policies normally entail voluntary or mandatory disclosure policies (ibid). Common examples include the toxics release inventory, lead paint disclosures, drinking water quality notices, and the International Standards Organization's voluntary ISO 14001 program. Empirically research into corporate disclosure has yielded mixed results. Jin and Leslie (2003) found that mandatory hygiene cards positively affected restaurant quality and health outcomes, while Delmas, Montes-Sancho & Shimshack (2009) found that mandatory disclosure over utility electricity generation mixes resulted in an increase in cleaner fuels. However, Lyon and Kim (2011) found that firms participating in Department of Energy's Voluntary Greenhouse Gas Registry engaged in 'green-washing', by selectively reporting emission reductions when overall firm emissions were increasing.

Information polices at the consumer level entail providing better information over the unobservable environmental impact of a consumer's behavior (Delmas & Lessem, 2011). This information can be feedback over their own behavior, social norms over aggregate behavior, or publicly disclosed information about a specific individual's behavior. In a number of studies in the electricity industry, improved feedback over an individual's own electricity usage has been shown to reduce electricity consumption by 4 to 12 percent on average (Darby, 2006; Abrahamse et al, 2005; Ehrhardt-Martinez et al, 2010), although many studies report finding no or perverse effects (Kihm et al, 2010: Klos et al, 2008; Allen & Janda, 2006; Sulyma et al, 2008; Sexton et al., 1987). Information over social norms has been shown to be effective at inducing conservation in a number of settings, including: recycling (Schultz, 1999), towel re-use (Goldstein, Cialdini, and Griskevicius, 2008), litter reduction (Cialdini, Reno, and Kallgren 1990), water conservation (Ferraro & Price, 2011) and energy conservation (Schultz et al, 2007; Ayers et al, 2009; Allcot, 2010; Costa & Kahn, 2011).

Eco-labels are the prime example of a product level information policy. The aim of eco-labels is to reduce the information asymmetry between producers and consumers that arises since consumers are not present during the production of the product and therefore cannot assess its environmental qualities. Attributes such as environmental quality, which cannot be verified before or after purchase, are called credence attributes (Darby & Karni, 1973). Credible eco-labels transform credence attributes into search attributes, where search attributes, such as color, size or price, can be identified by consumers prior to purchase (Nelson, 1970; Sammer & Wüstenhagen, 2006). The term eco-label commonly refers to a producer's right to use a symbol or phrase on their product labels, after passing a voluntary third-party environmental certification (Leire & Thidell, 2005; Rex & Bauman 2007). The International Standards Office (ISO) gives a broader description of eco-labels, classifying them as either mandatory or voluntary, with voluntary split into three types. The commonly used eco-label definition above would be categorized as Type I, whereas Type II are self-declared environmental claims and Type III are quantified environmental claims, usually to do with the lifecycle impact of the product.

The primary question that has occupied researchers over eco-labels, is whether consumers value eco-production and actually use it as a search attribute in purchasing products. Teisl et al (2002) found a premium for dolphin safe tuna using US supermarket scanner data, although identification is not clear since there is no cross sectional variation in certification. Looking at apparel catalogues, Nimon and Beghin (1999) found an eco-label premium for organic cotton clothing, but not for low impact dyes. Using a panel of weekly shopping data for Scandinavian consumers, Bjorner et al (2004) found that the Nordic Swan eco-label increased the probability of purchase for toilet paper and paper towels, but not detergents. In a study of eco-labeled hotels in Costa Rica, Rivera (2002) found that eco-labels generated a price premium for the top rated eco-hotels (based on a green leaf rating), but generated an eco-penalty for hotels with lower eco-ratings compared to uncertified hotels. In a discrete choice experiment, Sammer & Wüstenhagen (2006) found that Swiss consumers are willing to pay more for better energy efficiency ratings on washing machines.

In the market for wine several studies have investigated willingness to pay for eco-labeled wine, although these studies for the most part rely on surveys rather than experiments or real-world data. Loureiro (2003) surveyed consumers on their willingness to pay for eco-labeled Colorado wines and finds a 13 cent price premium on a $10 bottle of wine, over non-eco-labeled Colorado wine. Bazoche et al (2008) used wine auctions to elicit willingness to pay for eco-labeled wine in France interacted with an information treatment over pesticide usage in wine production. The authors found neither an eco-label premium nor an information effect, although their experiment was possibly limited by a lack of a proper wine control group (the eco-labeled wines were of a different brand and quality to the non-eco-labeled wines). Brugarolas et al (2009) surveyed Spanish consumers about whether they would pay a premium for organic wine and found that 75 percent would be willing. Similarly Forbes et al (2009) found that 73% of surveyed New Zealand consumers reported that they would choose an eco-labeled wine over a similar non-eco-labeled wine. Although in a follow up study Forbes et al (2011) found that consumers were less expressed less concern about the impact of wine production on the environment than food production in general.

7.3 A MODEL OF PRODUCT CHOICE WITH ECO-LABELS

Green products have been defined as "impure public goods" because they yield both public and private benefits (Cornes & Sandler, 1996; Ferraro, Uchida, & Conrad, 2005; Kotchen, 2006). Altruistic consumers, who care about the environment, may receive a good feeling or "warm glow" from engaging in environmentally friendly activities that contribute to this public good (Andreoni 1990). Such warm glow altruism has been shown to be a significant motivator of eco-consumption amongst environmentally minded consumers (Clarke et al, 2003; Kotchen & Moore, 2007; Kahn & Vaughn, 2009), with green consumption acting as a substitute for donations to environmental organization (Kotchen, 2005). On the private good side, consumers care about the quality of the product. Green products may offer quality advantages over their brown counterparts such as increased health benefits (Loureiro et al, 2001; Miles and Frewer, 2001; Yridoe et al, 2005), but they may also suffer from quality problems such as archaic production and farming techniques (Galarraga Gallastegui, 2002; Peattie & Crane, 2005).

Since green products (G) are credence goods; consumers cannot ascertain their environmental qualities during purchase or use. Some consumers who are concerned about the environment or their health may prefer these products, but in the absence of further information are unable are unable to determine their environmental qualities of the product at the time of purchase. A credible eco-label (L) can solve this problem by reducing this information assymetry between consumers and producers.

$$U_{ij} = L_j * (\gamma_i * G_j) + q_j - \alpha_i p_j \tag{1}$$

The utility of consumer i purchasing product j depends on both the quality (q_j) and price price (p_j) of the good, with α_i being a measure of price sensitivity for the individual. If the product is green (j G) and the consumer is an environmentalist ($\gamma_i > 0$), then the consumer will derive additional utility from it only if a credible eco-label is present ($L_j = 1$). Since product quality is unknown prior to consumption, consumers will make inferences over the product's quality from observable signals such as products price, label characteristics and the eco-label itself.

$$q_j = q + L_j * \beta G_j + f(p_j) + \delta O_j \tag{2}$$

Consumer perceptions of the virtues of green production, β, can be positive or negative, depending on the product. Price (p_j) may also act as a positive signal of quality (Milgrom & Roberts, 1986). We model this signal as diminishing with price, so that $f'(p_j) > 0$ and $f''(p_j) < 0$. We are interested in the case, where at low prices, the increased quality from a change in price outweighs the disutilty of having to pay the price. Thus we restrict the model so that $f'(p_j) > \alpha_i$. Finally other product attributes (vector JO) may influence the quality assessment, with vector δ entering in positively or negatively depending on the attribute and product. Substituting equation 2 into into our utility function in equation 1 we get:

$$U_{ij} = L_j * [(\gamma_i + \beta) * G_j] + f(p_j) - \alpha_i p_j + \delta O_j \qquad (3)$$

Equation 3 shows that the additional quality signal that an eco-label carries can further add to its appeal ($\beta > 0$) or diminish it ($\beta < 0$). If $\beta > 0$ then all consumers will prefer the eco-labeled green product to its otherwise identical brown counterpart. However, if $\beta \leq 0$ then only the subset of environmentally concerned consumers whose taste for eco-products is so strong that $\gamma_i > -\beta$ will prefer the green product. In the model above, we assumed that the quality signals act independantly of each other. However, there may be interactions between the various quality signals with the signals amplifying or diminishing each other. Since we are interested in eco-labels we restrict these interactions to those between eco-labels and other quality signals.

$$q_j = q + L_j * G_j \beta (1 + f(p_j) + \delta O_j) + f(p_j) + \delta O_j \qquad (4)$$

Substituting equation 4 into the utility function in equation 1 yields:

$$U_{ij} = L_j * G_j * (\gamma_i + \beta[1 + f(p_j) + \delta O_j]) + f(p_j) - \alpha_i p_j + \delta O_j \qquad (5)$$

As before, all consumers will prefer an eco-labeled green product over its brown counterpart when green production is viewed as being of a higher quality ($\beta < 0$). However, if $\beta \leq 0$ then the share of consumer who prefer the green product will decrease as price and other measures of quality increase. This is because consumers will only purchase the green product when $\gamma_i > -\beta[1 + f(p_j) + \delta O_j]$.

7.4 HYPOTHESES

The two models above give us a number of testable hypotheses. Since we are interested in the class of eco-products where eco-labels send a potentially negative quality signal, we restrict our hypotheses to those involving the cases where perceived quality is negative or zero.

1. H1: There is an absolute preference for eco-labeled products over non-eco-labeled products, which does not vary across price and other quality signals.
2. H2: There is a relative preference for eco-labeled products over non-eco-labeled products, which varies negatively with price and other quality signals, when eco-labeled products are perceived as lower quality.
3. H3: Environmentally minded consumers will prefer eco-labeled products over non-eco-labeled products.

Hypothesis 1 follows directly from equation 3 and says that for two similar products, those consumers whose environmental ideology exceeds the negative quality signal of the eco-label, will prefer the eco-labeled good regardless of the characteristics of the two products. Hypothesis 2 says that the share of consumers who prefer an eco-labeled product over an otherwise identical counterpart will vary with the characteristics of the product. In particular, the highest share of consumers will prefer the eco-labeled product when other quality signals are low. This is because they value the eco-labeled product's environmental attributes, and are unconcerned about the eco-label quality signal, since they have already inferred that quality is low. However as price and other quality signals increase, the share of consumers who prefer the eco-labeled product will decrease, since the low quality signal that the eco-label sends out is now more informative. This follows directly from equation 5.

We test these hypotheses by conducting an online discrete choice experiment, where consumers chose between eco-labeled and non-eco-labeled wines.

7.5 METHODOLOGY

To examine consumer preferences over eco-labels and other quality signals we ran an online discrete choice exercise, also known as a choice-based conjoint (CBC) exercise. CBC is a useful analytic technique for evincing consumer preferences in that it mirrors real-world choices as closely as possible, while still allowing the experimenter to randomize across prices and product attributes in a way that is not possible with real-world data. It also allows the experimenter to examine only those product attributes most relevant to the study. In our discrete choice exercise, consumers were shown images of four different wine bottle labels and asked to choose between them. They also had the option of choosing not to purchase any of the bottles on display, making the exercise more realistic (Louviere et al. 2000). Similar experiments on wine choice had variously examined the influence of medals (Lockshin et al, 2006), region of designation (Mtimet & Albisu, 2006) and back label information (Mueller et al, 2010). In addition to the CBC exercise, respondents also completed a survey which included demographic and attitudinal questions.

7.5.1 WINE LABELS AND WINE ATTRIBUTES

Wine labels are important in the wine purchase decision since the majority of wine purchases are unplanned, with consumers unaware of the quality difference between wines (Chaney, 2000; Bombrun & Sumner, 2003). Moreover, expert reviews that reveal wine quality are typically only available for the minority of wines at the top end of the price spectrum.2 We decided to focus our analysis on Californian wines produced for the US market. The United States is the largest wine consuming market in the world with retail sales totaling $30 billion in 2010 (Wine Institute, 2011). Californian wines dominate the US wine market, accounting for 90 percent of US production and 61 percent of US wine sales, by volume (ibid). The US wine market is an ideal backdrop to investigate the potentially negative effect of eco-labels, owing to potential quality concerns over eco-labeled wine and confusion over wine eco-labels.

Each wine bottle label in our choice set had five attributes: brandname, price, eco-label, region, and varietal. We created fictitious brands so that we did not need to worry about consumer knowledge and perceptions of existing brands. This was done by selecting names from a list of popular French lastnames. Four different brands were used, Chesnier, Challoner, Rutherfields, and Louis Devere, none of which corresponded to existing wineries. Four price levels were chosen, ranging from $8 to $29 in discrete $7 intervals.3 This range was chosen after a brief survey of the wine buying behavior of UCLA Anderson Business School faculty and students, and is higher than the $8 average selling price of a Californian wine in the US. Two Californian wine regions were used: the prestigious and well known Napa Valley and the lesser known and less-prestigious Lodi.To simplify the analysis, all bottles were of the same varietal - cabernet sauvignon. In 2009, cabernet sauvignon was the most popular Californian red wine varietal sold in the US (Wine Institute, 2010). We specifically chose a red wine to accentuate any potential eco-label quality concerns, for reasons that will be explained below.

The final attribute was eco-labels. Wines labels either had one of two eco-labels or nothing. These eco-labels are discussed below.

7.5.2 WINE ECO-LABELS

In the US wine industry, there are several competing eco-labels (see Appendix) related to environmental certification that are still not well recognized and understood by consumers (Delmas, 2008). Two of these labels are issued by the United States Department of Agriculture (USDA) and follow the U.S. National Organic farming standard, which prohibits the use of additives or alterations to the natural seed or plant, including, but not limited to, pesticides, chemicals, or genetic modification. The first of the USDA standards, "wine made from organically grown grapes", applies only to the production of the grapes, whereas the second, "organic wine", has prescriptions for the wine production process too. In particular, organic wine is prohibited from using sulfites in the wine-making process. Since sulfites help to preserve the wine, stabilize the flavor and eliminate unusual odors, wine produced without added sulfites may be of

lower quality (Waterhouse, 2007). Such quality concerns are most pertinent for red wines, which are usually kept for longer periods before consumption than white wines. This potential quality check does not apply to
wine made with organic grapes, which may add sulfites in the production
process. Other wine eco-labels include the internationally administered
"biodynamic" label and a variety of regional eco-labels, such as the "Lodi
Rules" label.

Eco-labeled wine provides a public good by engaging in environmentally friendly production practices that reduces the environmental degradation associated with conventional wine production, such as groundwater
depletion, water pollution, effluent run-off, toxicity of pesticides, fungi-
cides and herbicides, habitat destruction, and loss of natural biodiversity
(Warner, 2007).

7.5.3 WINE ATTRIBUTES AND QUALITY

On the private good side, the benefit to consumers of eco-labeled wine is
less clear. Wine made from organic grapes is free from pesticides and other
potentially harmful toxins, while organic wines do not add sulfites in production. Sulfites have long been associated with various health problems
such as asthma (Valley & Thompson 2001) and nasal blockages (Anderson et al 2009), and are also incorrectly blamed for causing wine-induced
headaches (Waterhouse, 2007). Research has shown that consumers do
view organic foods as healthier than conventional products (Loureiro et
al, 2001; Miles and Frewer, 2001; Yridoe et al, 2005), although they may
perceive there to be fewer health benefits from processed products such as
wine (Forbes et al, 2011).

Although eco-labeled wine may deliver some health advantages, consumers may perceive its main effect on the private aspect of consumption to be a reduction in quality. Quality concerns may arise for a number
of reasons. Firstly, organic wine, which is made without added sulfites,
may indeed be of a lower quality than conventionally produced wine. This
quality problem may incorrectly spillover to consumer perceptions of wine
made from organic grapes, if consumers are unaware of the distinctions

between the two labels. In our survey of 830 respondents, we find that although most are familiar with the concept of eco-labeled wines, 67 percent were unaware of the difference between the two labels. Quality concerns may also exist because early generations of eco-labeled wines, like many other eco-labeled products, were often experimental products, made by marginal producers and hence of variable quality (Cox, 2000; Galarraga Gallastegui, 2002; Peattie & Crane, 2005). This poor quality reputation may persist in the minds of consumer.

In the model of eco-label product choice we posited that price could act as a positive signal of product quality (following Milgrom & Roberts, 1986). A number of empirical wine demand studies support this supposition. Hedonic wine studies have found that quality, as assessed by professional wine reviewers, is a positive predictor of wine price (Bombrun & Sumner, 2003; Delmas & Grant, 2010; Landon & Smith, 1998). In study of wine choice in restaurants, where consumers were most likely unaware of wine quality, Durham et al (2004) found that demand increased with price for part of the price range. This was even after controlling for whether a wine was the lowest priced in its respective category. Similar results were obtained in discrete choice and experiments by Lockshin et al (2006) & Mtiment & Albisu (2006).

Region of origin (also known as appelation) has been shown to be a significant predictor of wine quality (Benjamin & Podolny, 1999). Of our two regions, Napa is known as a high quality producer and is the most famous location of wine production outside of Europe (Warner, 2007). Napa offers an ideal mixture of climate and soil conditions to produce a variety of premium varietals and is the oldest wine producing region in California (ibid). Lodi is less well known than Napa and has only been producing premium quality varietals for the last twenty years (ibid).

7.6 IMPLEMENTATION

Each experiment participant completed seven online discrete choice tasks and answered an online survey. The survey questions followed the discrete choice exercise so as to not bias the discrete choice responses. Since survey questions were focused on existing behaviors rather than attitudes, we feel it unlikely that participation in the discrete choice exercise caused bias in our survey results.

7.6.1 RECRUITMENT

Potential participants were asked to take part in an online survey related to wine preferences. Flyers advertising the survey were placed in several wine stores across the greater Los Angeles area and advertisements were placed on Facebook wine interest groups with membership totaling almost 100,000 people. Multiple emails were sent by both the authors and an undergraduate research team to professional and social contacts, with 4,845 people directly contacted. These primary contacts were asked to forward the survey to secondary contacts, although quantifiable information on the success of this strategy was not available to the authors. To motivate participation, a case of high quality wine was offered as a prize to a randomly drawn participant. Respondents were unable to take the survey more than once. The survey was taken by 1,142 participants and after removing foreign and incomplete entries, we were left with 883 valid responses. Although the majority of responses were centered in Los Angeles County (57 percent) and California (82 percent), the remaining respondents were drawn from 31 other US states.

As could be expected given the recruitment methodology, the experiment sample was over-represented by students relative to the general California population. This can be seen in Table 1, below. This results in a lower average age for the sample than the population. The experiment sample is also more educated and has higher incomes than the general population. This income-education bias is possibly alleviated somewhat in that the true wine buying population of California is possibly wealthier and better educated than the population average. Some support for this is given by a 2009 Gallup poll that showed that a small majority of college grads preferred wine over beer, whereas the vast majority of those who did not attend college preferred beer to wine (Gallup 2009). Lockshin et al (2006) report similar results for Australia.

Respondents report that on average they purchase organic products one out of every three trips to the grocery store, with 36 percent of respondents purchasing organic products on at least half of store visits. Similarly, about 20 percent of the sample reports being members of an environmental organization. By the nature of the population (young, students) respondents are probably more environmentally friendly or "greener" than average, although it should be noted that green consumerism is an increasingly important trend in the developed world. According to the Organization for Economic Co-operation and Development (OECD), "27% of consumers in OECD countries can be labeled 'green consumers' due to their strong willingness-to-pay and strong environmental activism" (OECD, 2005). In the U.S. retail sales of organic foods increased from US$3.8 billion in 1997 to US$26.6 billion in 2010 (Organic Trade Association, 2010). As an additional measure of environmentalism we linked each respondent to their state of resident's League of Conservation Voters (LCV) environmental rating for 2010.

7.6.2 DISCRETE CHOICE EXERCISE

Experiment participants were initially asked to complete seven choice tasks. In each choice task the respondent was asked to imagine that he/she was attending a seated dinner party with family and friends and needed to choose a bottle of wine to bring along for the occasion. Respondents were then presented with images of four different bottles of wine, each with a different price. The images were truncated to put focus on the wine bottle labels. Subjects were asked to choose which bottle of wine they would purchase, with the option of choosing to purchase none of them. Respondents selected their prefered option by clicking on it. An example of a choice task is shown in figure 1 below.

Increasing the number of choice tasks faced would have helped to better identify interactions between wine attributes. However, this would have come at the cost of greater attrition, especially since the respondents were unpaid volunteers. Instead, we offered four different versions of the survey, each with its own seven choice tasks and unique attribute

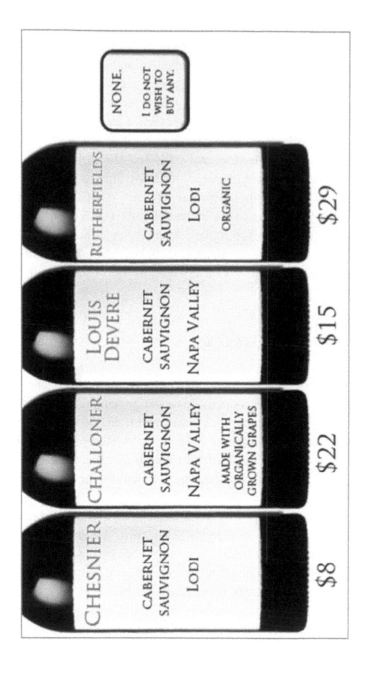

FIGURE 1: Wine Choice Tasks

combinations. This has the same effect as increasing the number of choice tasks (after controlling for individual attributes).

Each bottle of wine had one level of each of the five attributes. The levels of the attributes were randomized across the 28 different choice tasks (4x7) using Sawtooth Software's Choice-Based Conjoint Software. An algorithm was used to insure each level of each attribute appeared an equal number of times across all surveys, but did not repeat in the wine bottles within each choice task. This was done to make sure that the respondent did not see the same level, (e.g., the same price) across all the choices in one task. To ensure that the choice set was not dominated by eco-label wines, we doubled the number of non-eco-labeled wines. Thus every choice set had one organic wine, one made with organic grapes wine and two non-eco-labeled wines. Table II, shows each of the attribute levels and its display and selection frequency.

7.7 ECONOMETRIC SPECIFICATION

Each subject was given 7 discrete choice tasks to complete ($C \in 1..7$]). In each task the subject was asked to choose between hypthetically purchasing one of 4 different bottles of wine, or buying none of them. Each bottle of wine is respresented by a vector of attributes $W_j^c \in [0..4,]$, $C \in [A..G]$, where $j = 0$ indicates the none option. No bottles of wine were repeated for a given consumer. The ordering of the discrete choice tasks were randomized across consumers, although within a given choice task the four bottles always appear in the same order (which resulted from an initial randomization). Individual attributes were obtained from the survey and are represented by vector , X_i, $i \in [1..N]$. The interaction between subject and product attributes is $Z_{ij}^C = \text{vec}[W_j^{c'} X_i]'$. The outcome variable y_{ij}^C, is a dummy variable indicating whether the bottle was purchased or not.

The utility subject i gets from bottle j is:

$$U_{ij} = X_i B_X' + W_j^C B_W' + Z_{ij}^C B_Z' + \varepsilon_{ij}^C = V_{ij}^{C,V} + \varepsilon_{ij}^C$$

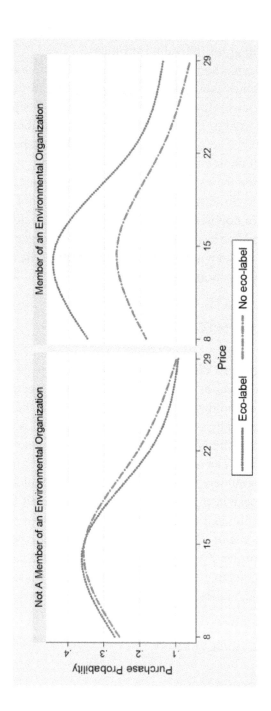

FIGURE 2: Absolute preferences over eco-labeled wine vs. non-eco-labeled wine.

FIGURE 3: Relative preference over eco-labeled verse non-eco-labeled wine

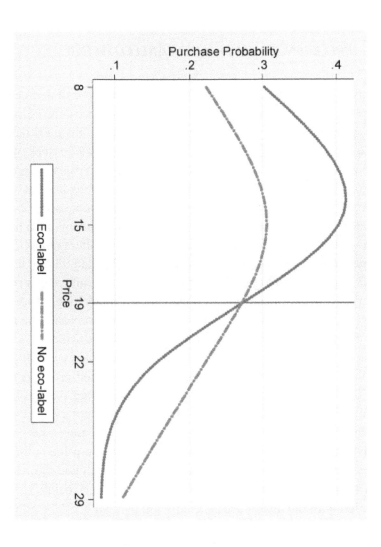

To account for repeated choice tasks by each subject, we cluster standard errors at the subject level.

7.8 RESULTS

7.8.1 PREFERENCES OVER ECO-LABELED WINE

In Table III we examine preferences for eco-labeled and non-eco-labeled wine. All comparisons are with respect to a non-eco-labeled wine from Lodi, priced at $8. Equation 1 examines whether consumers have an absolute preference for eco-labeled over non-eco-labeled wines, where this preference does not vary with other wine characteristics (hypothesis 1). Consumers are slightly more likely to purchase eco-labeled wine (2.4 percentage points), but this difference disappears in equation 2, when we include individual level controls. Only those respondents who buy a high proportion of organic already, and/or who are members of environmental organizations prefer eco-labeled over non-eco-labeled wine. This is illustrated graphically in Figure 2. In both equations we can see that consumers prefer wine from Napa and wine priced at $15. Interestingly, those respondents who are wealthier, better educated, spend more on wine and drink wine frequently are less likely to choose eco-labeled wines.

In equation 3, we allow the eco-label to interact with other quality signals to see whether the preference for eco-labeled products decreases with other quality signals (hypothesis 2). We interact the eco-label with a dummy for if the price is high (price=22 or price=29) and a dummy for Napa. We find that consumers are 14.6 percent more likely to buy an eco-labeled than non-eco-labeled product when the price is low and the wine is from Lodi. However, when the price is high and the wine comes from Napa, this preference reverse, with consumers being 13.1 percent more likely to buy a non-eco-labeled wine over an eco-labeled wine. The same results hold when we include individual characteristics in equation 4. All of the individual characteristics have the same sign and magnitude as in equation 2. The price-penalty of eco-labels is illustrated graphically in Figure 3.

These results indicate that respondents obtain some warm glow value from eco-labeled wine, but also interpret it as a signal of low quality. If respondents made no inferences over wine quality, they would always prefer an eco-labeled wine over an otherwise identical non-eco-labeled wine, regardless of other attributes. Instead we find that preferences over eco-labeled wine vary with these other attributes. One interpretation of the data is that when respondents have already inferred that a wine is low quality from price and other attributes, then the additional low quality signal from the eco-label is unimportant, and respondents receive just the warm glow of eco-consumption. However, as price and other quality signals increase, the eco-labels quality signal becomes more pertinent and outweighs the warm-glow of eco-consumption, shifting preferences towards non-eco-labeled wine. These findings support hypothesis two over hypothesis one, since preference over eco-labels in indeed relative rather than absolute.

7.8.2 KNOWLEDGE AND PREFERENCES OVER ECO-LABEL TYPES

In table IV we examine consumer preferences over the two different USDA eco-labels; organic and made with organic grapes. Organic wine undergoes a different production process to non-organic wine often resulting in inferior quality. The same is not true of wine made with organic grapes. Thus it should not suffer from the same quality penalty as organic wines. Again comparisons are made with respect to a non-eco-labeled wine from Lodi, priced at $8. We exclude individual controls in this specification because our sample size does not permit us to include separate individual coefficients for each eco-label type in addition to the increased number of eco-labels and interactions.7

Equation 1 shows that there is no difference in how consumers evaluate organic wine and wine made from organic grapes. Apart from the coefficient on eco-label interacted with Napa, there are no significant differences between consumer preferences for the two different eco-labels. This indicates that quality concerns about organic wine are potentially being incorrectly attributed to wine made with organic grapes. Presumably consumers who are aware of the difference between the two eco-labels would not make this mistake. We examine this in regression 2, by estimating separate coefficients for the eco-label price penalty for those who are informed of the difference and those who are not. Regression 2 shows that those who are informed do not on place as high a premium on low priced eco-labeled wine as uninformed respondents, but value the two eco-labels equally.8 More importantly, informed consumers do place a price penalty on organic wine, but no price penalty on wine made from organic grapes.9 Uninformed consumers place an identical price penalty on both eco-labels. This is illustrated graphically in Figure 4.

We interpret these results as signifying that consumers who know that wine made with organic grapes is produced in the same fashion as non-eco-labeled wine do not put a quality penalty on wine made with organic grapes. Interestingly, despite the superior quality input of organic grapes, these consumers do place an absolute premium on this wine. On the other hand, consumers who are unaware of this difference tend to treat organic wine and wine made with organic grapes similarly, imposing a perceived quality penalty on both.

7.8.3 REPUTATION AND BRAND NAME

In our discrete choice exercise we used fictitious wine brands, so that our results would not be conflated by existing consumer brand beliefs. However brand names may be a strong quality signal that can overcome the quality-tradeoff that eco-labels present. Some insight into this question can be gained by looking at the effects of region, which can act as meta-brand for wines, signifying common quality levels for all producers (Benjamin & Podolny, 1999). If a quality brand was sufficient to overcome quality concerns for eco-labeled wines, then we would expect the interaction between brand and eco-label to be weakly positive. However, when

FIGURE 4: Relative preference over eco-label

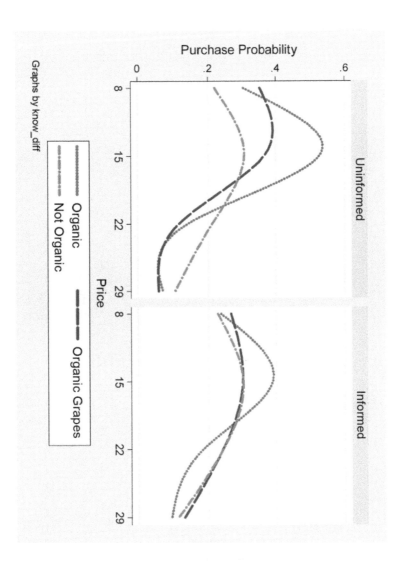

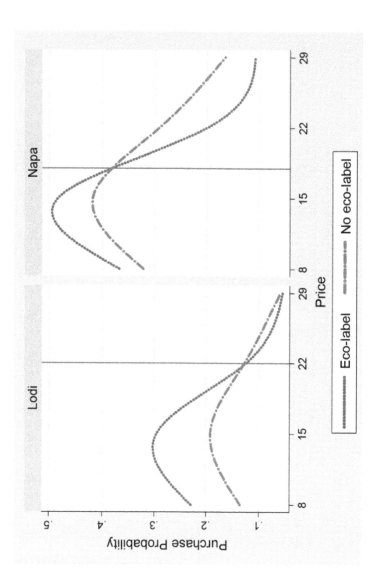

FIGURE 5: Relative preference over eco-labeled verse non-eco-labeled wine by region

we examine the interaction between the meta-brand Napa and eco-labels in Table III, we find a negative coefficient, implying that the high quality signal given by region is insufficient to over-ride the low quality signal given by eco-label. Individual winemakers seem to be aware of this, with a number of high quality wine makers producing eco-certified wine, without labeling it as such (Delmas & Grant). Relative demand by region is displayed graphically in Table V.

7.9 DISCUSSION AND CONCLUSION

Eco-labels are often developed by government agencies and non-governmental organizations which are separate to the industries that produce and sell the eco-product. The goal of these agencies is to reduce the information asymmetry between producers and consumers over the environmental attributes of a good. If an eco-label is effective it will command a premium amongst environmentally minded consumers and thus allow manufacturers to recoup the additional costs of cleaner manufacturing practices. However, by focusing on the information asymmetry between producers and consumers, rather than how the label meets consumer needs, agencies may develop eco-labels that send an irrelevant, confusing or detrimental message to consumers.

We develop a model where consumers receive a "warm glow" from eco-consumption, but also view the eco-label as a signal of low quality. We tested this empirically with an online discrete choice experiment focused on choices over eco-labeled wines. In the US there are two government certified eco-labels for wine. One label is associated with potentially low quality production techniques, while the other is not. The majority of the 830 participants in our experiment were unaware of the difference between these two labels. We found that respondents preferred eco-labeled wines over an otherwise identical counterpart, when the price was low and the wine was from a low quality region. However these preferences were reversed if the wine was expensive and from a high quality region. These results indicate that respondents obtain some warm glow value from

eco-labeled wine, but also interpret it as a signal of low quality. If respondents made no inferences over wine quality, they would always prefer an eco-labeled wine over an otherwise identical non-eco-labeled wine,
regardless of other attributes. One interpretation of these results is that
when respondents have already inferred that a wine is low quality from
price and other attributes, then the additional low quality signal from the
eco-label is unimportant, and respondents receive just the warm glow of
eco-consumption. However, as price and other quality signals increase,
the eco-labels quality signal becomes more pertinent and outweighs the
warm-glow of eco-consumption, shifting preferences towards non-eco-
labeled wine. This eco-quality penalty holds for both types of eco-labels,
even though it should only apply to the eco-label associated with quality
concerns. However, for the minority of respondents who were informed of
the difference between the two eco-labels, the eco-quality penalty applied
only to the potentially low quality label. This suggests a potential solution
to the problem, in the form of consumer education.

By ignoring potential quality signals from eco-labels, the market for
eco-labeled wine has been severely limited. Government certified eco-
labeled wine obtained just 0.1% of the overall wine market in 2009, compared with 3.5% for the overall market for similar government eco-labels
for other eco-labeled products (Wine Institute, 2010). This lack of market
penetration combined with consumer confusion has opened up the door to
a number of other unregulated eco-labels, which may be less green than
government certified eco-labels. These eco-labels may create further confusion and erode credibility in the eco-wine market.

The lessons from the wine industry for other eco-labeling initiatives
are clear. An eco-label premium is essential for an eco-industry to sustainably exist. Thus any eco-labeling initiative needs to ensure that it will
deliver such premiums. Focusing purely on information asymmetries will
not necessarily create eco-labels that align eco-products with the needs
of consumers. Instead government organizations need to work with producers and marketers to ensure that eco-labels provide information that
clearly communicate their value proposition to consumers, without creating further confusion, or additional unintended product signals.

REFERENCES

1. Abrahamse, W., Steg, L., Vlek, C., & T. Rothengatter (2005): "A Review of Intervention Studies Aimed at Household Energy Consumption", Journal of Environmental Psychology, Vol. 25, pp.273-291.

2. Allcott, H. (2011): "Social norms and energy conservation", Journal of Public Economics, Vol. 95, Issues 9-10, pp.1082-1095

3. Allen, D., & K. Janda, (2006): "The Effects of Household Characteristics and Energy Use Consciousness on the Effectiveness of Real-Time Energy Use Feedback: A Pilot Study Continuous Feedback: The Next Step In Residential Energy Conservation?" 2006 ACEEE Summer Study on Energy Efficiency in Buildings, pp.1-12.

4. Anderson, M., Cervin-Hoberg C. & L Greiff (2009): "Wine produced by ecological methods produces relatively little nasal blockage in wine-sensitive subjects", Acta Oto-laryngologica 129:11,

5. Andreoni, J. (1990): "Impure Altruism and Donations to Public Goods: A Theory of Warm-Glow Giving.", Economic Journal, vol. 100, pp. 464-477.

6. Ayers, I., Raseman, S. & A. Shih (2009): "Evidence from Two Large Field Experiments that Peer Comparison Feedback Can Reduce Residential Energy Usage", NBER Working Paper 15386.

7. Bazoche P., Deola C. & L.G. Soler (2008): "An experimental study of wine consumers' willingness to pay for environmental characteristics", 12th Congress of the European Association of Agricultural Economists – EAAE 2008

8. Benjamin, B. & J. Podolny (1999): "Status, Quality, and Social Order in the California Wine Industry", Administrative Science Quarterly, vol. 44, pp.563-589

9. Bhaskaran, S., Polonsky, M., Cary, J. & S. Fernandez (2006): "Environmentally sustainable food production and marketing. Opportunity or hype? British Food Journal, Vol. 108 No. 8, pp. 677-690

10. Bjorner T., Hansen, L. & C. Russell (2004): "Environmental labeling and consumers' choice—an empirical analysis of the effect of the Nordic Swan", Journal of Environmental Economics and Management, vol. 47, pp.411–434

11. Bombrun, H., & D. Sumner (2001): "What determines the price of wine?" AIC Issues Brief, 18, pp. 1-6.

12. Brugarolas, M., Martinez-Carrasco, L., Bernabeu, R. & A. Martinez-Poveda (2009): "A contingent valuation analysis to determine profitability of establishing local organic wine markets in Spain", Renewable Agriculture and Food Systems, Vol. 25(1), pp.35–44

13. Chaney, I. M. (2000): "External search effort for wine", International Journal of Wine Marketing, 12(2), 5–21.

14. Cialdini, R.B., Kallgren, C.A. & R.R. Reno (1990): "A Focus Theory of Normative Conduct: Recycling the Concept of Norms to Reduce Littering in Public Places", Journal of Personality and Social Psychology, Vol. 58 no. 6, pp. 1015-1026.

15. Clarke C., Kotchen M. & M. Moore (2003): "Internal and external influences on pro-environmental behavior: Participation in a green electricity program", Journal of Environmental Psychology, vol. 23

16. Cornes, R., & Sandler, T. (1996): The theory of externalities, public goods, and club goods (2nd ed.), Cambridge, UK: Cambridge University Press.

17. Costa, D. L., & Kahn, M. E. (2010): "Energy Conservation "Nudges" and Environmental Ideology: Evidence from a Randomized Field Experiment", NBER Working Paper 15939

18. Cox J. (2000): "Organic wine growing goes mainstream", winenews.com available at: http://www.thewinenews.com/augsep00/cover.html (last accessed 12 December 2011)

19. Crespi, J. M., & Marette, S. (2005): "Eco-labelling economics: Is public involvement necessary?" In S. Krarup & C. S. Russell (Eds.), Environment, information and consumer behavior (pp. 93-110). Northampton, MA: Edward Elgar.

20. Darby, S. (2006): "The effectiveness of feedback on energy consumption", Working Paper, Oxford Environmental Change Institute (2006) (April).

21. Darby M, & E. Karni (1973): "Free Competition and the Optimal amount of Fraud", Journal of Law and Economics, Vol. 16:, pp.67–88.

22. De Boer, J. (2003): "Sustainability Labeling Schemes: The logic of their claims and their functions for stakeholders", Business Strategy and the Environment, vol. 12, pp.254–264

23. Delmas, M. (2008): "Perception of eco-labels: Organic and biodynamic wines" (Working Paper). Los Angeles: UCLA Institute of the Environment.

24. Delmas M. & L. Grant (2010): "Eco-labeling Strategies and Price-Premium: The Wine Industry Puzzle", Business and Society

25. Delmas M. & N. Lessem (2011): "Saving Power to Conserve Your Reputation? The Effectiveness of Public versus Private Information", (Working Paper). Los Angeles: UCLA Institute of the Environment.

26. Delmas, M., M.J. Montes-Sancho & J.P. Shimshack (2009): "Information Disclosure Policies: Evidence from the Electricity Industry", Economic Inquiry, Volume 48, Issue 2, pp. 483–498.

27. Dietz, T. & P. Stern (2002): "Exploring New Tools for Environmental Protection", in T Dietz, & P. Stern (Eds.) New Tools for Environmental Protection: Education, information and Voluntary Measures, National Academies Press

28. D'Souza, C. Taghian, M., Lamb, P. & R. Peretiatko (2006): "Green decisions demographics and consumer understanding of environmental labels", International Journal of Consumer Studies, vol. 31, pp.371-376

29. Durham, C., Pardoe, I. & E. Vega (2004): "A Methodology for Evaluating How Product Characteristics Impact Choice in Retail Settings with Many Zero Observations: An Application to Restaurant Wine Purchase" Journal of Agricultural and Resource Economics, Vol. 29, No. 1, pp. 112-131

30. Ehrhardt-Martinez, K., Donelly, K.A. & J.A. Laitner (2010): "Advanced Metering Initiatives and Residential Feedback Programs: A Meta-Review for Household Electricity-Savings Opportunities", American Council for an Energy Efficient Economy (ACEEE), report number E105.

31. Ferraro, P.J. & M.K. Price (2011): "Using Non-Pecuniary Strategies to Influence Behavior: Evidence from a Large Scale Field Experiment", NBER working paper 17189

32. Ferraro, P., Uchida, T., & J.M. Conrad (2005): "Price premiums for eco-friendly commodities: Are "green" markets the best way to protect endangered ecosystems?", Environmental and Resource Economics, 32, 419-438.

33. Forbes, S., Cullen, R., Cohen, D., Wratten, S. & Joanna Fountain (2009): "Consumer attitudes regarding environmentally sustainable wine: an exploratory study of the New Zealand marketplace", Journal of Cleaner Production, Vol. 17, pp.1195–1199

34. Forbes, S., Cullen, R., Cohen, D., Wratten, S. & Joanna Fountain (2011): "Food and Wine Production Practices: An Analysis of Consumer Views", Journal of Wine Research, Vol. 22:1, pp.79-86

35. Galarraga Gallastegui, I. (2002): "The use of eco-labels: A review of the literature", European Environment, 12, 316-331.

36. Gallup (2009): "Drinking Habits Steady Amid Recession", http://www.gallup.com/ (last accessed 18 April, 2011)

37. Goldstein, N., Cialdini, R. & V. Griskevicius (2008): "A Room with a Viewpoint: Using Norm-Based Appeals to Motivate Conservation Behaviors in a Hotel Setting", Journal of Consumer Research, Vol. 35, pp. 472-482.

38. Jin, G., and P. Leslie (2003): "The Effect of Information on Product Quality: Evidence from RestaurantHygiene Grade Cards" Quarterly Journal of Economics, 118, pp. 409–51.

39. Kahn M. & R. Vaughn (2009): "Green Market Geography: The Spatial Clustering of Hybrid Vehicles and LEED Registered Buildings", The B.E. Journal of Economic Analysis & Policy, vol.9 issue 2, article 2

40. Kihm, S., Koski, K. & A. Mendyk (2010): "Focus on Energy – PowerCost Monitor Study", Energy Center of Wisconsin, Report Number 253-1

41. Klos, M., Erickson, J., Bryant, E. & S.L. Ringhof (2008): "Communicating Thermostats for Residential Time-of-Use Rates: They Do Make a Difference", ACEEE Summer Study on Energy Efficiency in Buildings.

42. Kotchen, M. J. (2005): "Impure public goods and the comparative statics of environmentally friendly consumption", Journal of Environmental Economics and Management, 49, 281-300.

43. Kotchen, M. J. (2006): "Green markets and private provision of public goods", Journal of Political Economy, 114, 816-845

44. Kotchen, M. & M.R. Moore (2007): "Private provision of environmental public goods: Household participation in green-electricity programs", Journal of Environmental Economics and Management, September 2007. 53, pp. 1-16

45. Landon S. & C.E. Smith (1998): "Quality Expectations, Reputation, and Price" Southern Economic Journal, Vol. 64, No. 3, pp. 628-647.

46. Leire, C., & A. Thidell (2005): "Product-related environmental information to guide consumer purchases—A review and analysis of research on perceptions, understanding and use among Nordic consumers", Journal of Cleaner Production, 13, 1061-1070..

47. Lockshin L., Jarvis W., d_Hauteville F., & J. Perrouty (2006): "Using simulations from discrete choice experiments to measure consumer sensitivity to brand, region, price, and awards in wine choice", Food Quality and Preference, vol. 17, pp. 166–178

48. Loureiro, M.L., McCluskey, J.J. & R. Mittelhammer (2001): "Assessing Consumer Preferences for Organic, Eco-labeled, and Regular Apples", Journal of Agricultural and Resource Economics, Vol. 26(2), pp.404-416

49. Loureiro, M. L. (2003): "Rethinking new wines: Implications of local and environmentally friendly labels", Food Policy, 28, 547-560.

50. Louviere, J. J., Hensher, D. A., & Swait, J. (2000): "Stated choice methods analysis and application", Cambridge: Cambridge University Press

51. Lyon, T.P. & E. Kim, (2011): "Strategic Environmental Disclosure: Evidence from the DOE's Voluntary Greenhouse Gas Registry." Journal of Environmental Economics and Management, vol. 61, pp.311-326.

52. Miles, S., & L. Frewer (2001):" Investigating specific concerns about different food hazards", Food Quality and Preference, vol. 12, pp. 47-61.

53. Milgrom, P. & J. Roberts (1986): "Price and Advertising Signals of Product Quality," Journal of Political Economy, University of Chicago Press, vol. 94(4), pp.796-821.

54. Mtimet, N. & L.M. Albisu (2006): "Spanish Wine Consumer Behavior: A Choice Experiment Approach", Agribusiness, Vol. 22 (3), pp.343–362

55. Mueller, S., Lockshin, L., Saltman, Y. & J. Blanford (2010): "Message on a bottle: The relative influence of wine back label information on wine choice", Food Quality and Preference, vol. 21 , pp.22–32

56. Nelson P. (1970): "Information and consumer behavior" Journal of Political Economy, Vol. 78, pp. 311– 329.

57. Nilsson, H. Tuncer, B. & A Thidell (2004): "The use of eco-labeling like initiatives on food products to promote quality assurance—is there enough credibility?" Journal of Cleaner Production, vol. 12, pp. 517–526

58. Nimon, W. & J. Beghin (1999): "Are Eco-Labels Valuable? Evidence from the Apparel Industry", American Journal of Agricultural Economics, Vol. 81, No. 4, pp. 801-811.

59. OECD (2005): "Effects of eco-labeling schemes: compilation of recent studies", Joint Working Party on Trade and Environment, paper # COM/ENV/TD(2004)34/ FINAL

60. Organic Trade Association (2011): "The OTA 2011 Organic Industry Survey" Available from http://www.ota.com/organic/mt.html

61. Peattie, K. (2001): "Golden Goose or Wild Goose? The hunt for the green consumer", Business Strategy and the Environment, vol. 10, pp.187-199

62. Peattie, K., & A. Crane (2005): "Green marketing: Legend, myth, farce or prophesy?", Qualitative Market Research: An International Journal, 8, 357-370.

63. Rex. E & H. Baumann (2007): "Beyond ecolabels: what green marketing can learn from conventional marketing", Journal of Cleaner Production, vol. 15, pp. 567-576

64. Rivera, J. (2002): "Assessing a voluntary environmental initiative in the developing world: The Costa Rican Certification for Sustainable Tourism", Policy Sciences, vol. 35, pp.333-360

65. Sammer K. & R. Wüstenhagen (2006): "The Influence of Eco-Labelling on Consumer Behaviour – Results of a Discrete Choice Analysis for Washing Machines", Business Strategy and the Environment, Vo. 15, pp.185-199.

66. Schultz, P. W. (1999): "Changing Behavior with Normative Feedback Interventions: A Field Experiment on Curbside Recycling", Basic and Applied Social Psychology, 21:1, pp.25-36

67. Schultz, P. W., Nolan, J. M., Cialdini, R. B., Goldstein, N. J., & V. Griskevicius (2007): "The constructive, destructive, and reconstructive power of social norms", Psychological Science, Vol. 18(5), pp. 429-34.

68. Sexton, R.J, Brown Johnson, N. & A Konakayama (1987): "Consumer Response to Continuous-Display Electricity-Use Monitors in a Time- of-Use Pricing Experiment", The Journal of Consumer Research, Vol. 14, No. 1, pp. 55-62.

69. Stern, P. (1999): "Information, Incentives, and Proenvironmental Consumer Behavior", Journal of Consumer Policy, vol. 22, pp. 461–478

70. Sulyma, I., Tiedemann, K., Pedersen, M., Rebman M.& M. Yu (2008): "Experimental Evidence: A Residential Time of Use Pilot", 2008 ACEEE Summer Study on Energy Efficiency in Buildings.

71. Teisl, M., Roe, B. & R. Hicks (2002) : "Can Eco-Labels Tune a Market? Evidence from Dolphin-Safe Labeling", Journal of Environmental Economics and Management, vol. 43, pp.339-359

72. Teisl, M., Rubin, J., Smith, M., Noblet, C., Cayting, L., Morrill, M., Brown, T. & S. Jones (2004): "Designing Effective Environmental Labels for Passenger Vehicle Sales in Maine: Results of Focus Group Research", Maine Agricultural and Forest Experiment Station Miscellaneous Report 434

73. Valley H. & P.J. Thompson (2001): "Role of sulfite additives in wine induced asthma: single dose and cumulative dose studies", Thorax vol. 56: doi:10.1136/thorax.56.10.763

74. van Amstel, M., Driessen, P. & P. Glasbergen (2008): "Eco-labeling and information asymmetry: a comparison of five eco-labels in the Netherlands", Journal of Cleaner Production, vol. 16, pp. 263-276

75. Warner, K.D. (2007): "The quality of sustainability: Agroecological partnerships and the geographic branding of California winegrapes", Journal of Rural Studies, vol. 23, pp.142–155

76. Waterhouse, A. (2007): "Sulfites in Wine", available online at http://waterhouse.ucdavis.edu/winecomp/so2.htm (last accessed 12 December 2011).

77. Wine Institute (2010): "California Wine Industry Statistical Highlights", available at http://www.wineinstitute.org/files/EIR%20Flyer%202008.pdf (last accessed 18 March 2011).

78. Wine Institute (2011): "2010 California/US Wine Sales", available at http://www.wineinstitute.org/resources/statistics/article584 (last accessed 12 December 2011).

79. Yridoe, E., Bonti-Ankomah, S. & R. Martin "Comparison of consumer perceptions toward organic versus conventionally produced foods: a review and update of the literature", Renewable Agriculture and Food Systems, vol. 20(4); pp.193–205

There are several supplemental files that are not available in this version of the article. To view this additional information, please use the citation on the first page of this chapter.

CHAPTER 8

DETERMINANTS OF WILLINGNESS-TO-PAY FOR SUSTAINABLE WINE: EVIDENCE FROM EXPERIMENTAL AUCTIONS

RICCARDO VECCHIO

8.1 INTRODUCTION

According to the review performed by Christ and Burritt (2013) key areas of environmental concern currently facing the global wine industry are: water use and quality issues, the production and management of organic and inorganic solid waste streams, energy use and the generation of greenhouse gas emissions, the use and management of chemicals in the vineyard and winery, land use issues and the impact on ecosystems. Indeed, like other food industries, the wine business has been increasingly impelled by market and regulatory drivers to assess, reduce and communicate environmental and social performances, particularly in certain countries with a shorter tradition in winemaking (Australia, New Zealand, the USA and South Africa). In addition, wine companies have realized that sustainability constitutes a means of differentiation, which is crucial for

Determinants of Willingness-to-Pay for Sustainable Wine: Evidence From Experimental Auctions . © Vecchio R. Wine Economics and Policy, *2,2 (2013), DOI: 10.1016/j.wep.2013.11.002. Open Access funded by UniCeSV, University of Florence. Used with permission from the authors*

increasing productivity and competitiveness. Consequently, sustainability has developed into a priority in the wine supply chain (Forbes et al., 2009 and Gabzdylova et al., 2009).

Despite the above-described scenario, the reasons behind consumers' adoption of sustainable practices, attitudes and intention to purchase sustainable wines remain largely unexplored (Barber et al., 2010). Furthermore, while many authors believe that consumers will not be willing to trade the quality of a wine off against environmental/social features (Lockshin and Corsi, 2012) – thus sustainable wines should be sold at the same price as regular wines – other scholars hold that sustainability is most likely to become a considerable competitive advantage in the international arena (Pullman et al., 2010 and Forbes et al., 2009). A major drawback of most of the published articles on sustainable wine is the use of contingent valuation techniques that do not capture actual behavior due to strong hypothetical bias. Indeed, unconstrained survey responses eliciting purchase intention, attitudes or product liking, used in most previous research on consumer valuation of ethical behavior, has been criticized for social desirability bias (Auger and Devinney, 2007) and the attitude–behavior gap (Carrington et al., 2010). To reduce such potential bias prominent authors recommend using specific products and incentive-compatible research methods. As previously demonstrated, auctions seem to be an effective method to obtain valid information on the perceived value of an attribute tested in the presence of external information; allowing one to know the monetary value attributed to a given label, brand or product while taking into account the economic constraint faced by the consumers (Lange et al., 2002). Nevertheless, this methodology has quite rarely been applied to wine (e.g. Combris et al., 2009; Lange et al., 2002, Combris et al., 2009 and Sáenz-Navajas et al., 2013).

This paper draws on experimental auctions conducted in Naples (Italy) to analyze the true value attached by consumers to social and environmental claims concerning wine. In particular, the research was designed to cast light on the importance of social/ethical and environmental attributes for young adult wine drinkers (i.e. individuals consuming wine at least once a month).

The remainder of this paper is arranged as follows. The next section discusses the importance of young individuals in today's wine market.

Subsequently a detailed description of the data gathering process and methods used is offered. The results of econometric analysis are then presented. Finally, our findings are discussed and compared with recent key studies, and future research avenues are outlined.

8.2 RESEARCH BACKGROUND

The wine market has experienced a huge change in geography of consumption over the last 40 years: a substantial share of total consumption has moved from large producing countries to those with a limited domestic production or none at all. Starting from the second half of the 1970s wine consumption has continuously decreased in traditional large producers in Europe and in South America (countries which used to be key consumers), and with the crisis in the Soviet Union there were sharp declines also in East and Central Europe. Meanwhile, however, starting from the 1960s consumption began to increase in Northern Europe, North America and Japan, countries which can now be considered traditional importers, and later, from the mid 1990s in countries which until that time were marginally involved with wine, namely Asia or non-producing countries in Central and South America; there was also a return to consumption in Central and Eastern Europe (Mariani et al., 2011 and Mariani et al., 2012). In traditional producing countries the decline in domestic consumption has been considered as an inevitable consequence of lifestyle changes and the wine industry has reacted by increasing its export propensity, reaching countries with a growing interest in wine (Rabobank, 2003 and Anderson and Nelgen, 2011). Increased competition on international markets, however, is inducing wine industry stakeholders of the main producing countries to identify national strategies to stabilize domestic wine consumption in terms of quantity and, if possible, to increase sales. In this perspective we can frame the initiative of the Argentine parliament in the spring of 2013, which declared wine as the <national beverage> and that of the Spanish parliament which, in the same period, formed the group called Asociación Parlamentaria por la Cultura de la Viña y el Vino (APCVV) to exploit the importance of wine as a core element of Mediterranean culture.

For about two years in Italy, the main association of wine producers, the Italian Wine Union, has been encouraging academics and policy makers to study the characteristics and expectations of Italian wine consumers with at least the same care which is applied to foreign consumers. Wine consumption in Italy started to decrease on a nationwide basis in the early 1970s. In those years, domestic consumption reached 60 million hectoliters, which corresponded to a per capita consumption of more than 1,001 per year, while at the end of the first decade of the new century, consumption had stabilized at just above 20 million hectoliters, with an annual consumption of less than 401 per capita. Reflecting a change in consumer behavior, the total amount of wine consumed by each individual has decreased but there has also been a decline in the total number of wine drinkers in Italy. The proportion of wine drinkers in the Italian population in the early 1990s was just under 60% while by 2010 it had fallen to just over 53%.

Detailed analysis of the contribution of different age classes to the change in consumption patterns shows that in recent years (2003–2010) the older age cohorts have largely contributed to this decline. Larger shares of young adults under 24 are becoming wine drinkers compared to individuals between 25 and 34 years. Even if among these young adults the proportion of daily drinkers continuously decreases – the share of occasional drinkers is increasing – wine is a product that tends to take root in the lifestyle of these individuals. Moreover, recent surveys (ISPO, 2012) indicate that in young Italians there prevails, unlike in France, Germany and the United Kingdom, a tendency to drink alcoholic beverages responsibly, and this would appear to reward wine consumption.

From a marketing perspective it is therefore extremely important to exploit these signals and strengthen the relationship between wine and the younger generation in order to bring about conditions for consumption growth. To achieve this it is of paramount importance to characterize the image of the product consistently with the issues to which the younger generation appears to be more sensitive.

In the last decade in many producer countries, the wine industry has devoted considerable resources to the identification of pathways to adapt production processes to the principles of sustainable development and the establishment of protocols for the evaluation of sustainability behavior. This is also happening in Italy and it is consequently interesting to ascertain to what extent the issue of sustainability can be useful to contribute to the embedding of wine consumption among the younger generation. Sustainability appears to be a potentially useful issue as younger generations seem to be particularly sensitive to this topic (UNEP, 2011). It has been demonstrated that the so-called Millennials care more compared to other cohorts about the environmental impact of the wine industry (MacDonald et al., 2013 and Thach and Olsen, 2006). On the other hand, interest in sustainability does not automatically translate into purchases of sustainable food (Vermeir and Verbeke, 2006), as other factors strongly impact behavior.

As a result, in order to understand to what extent the issue of sustainability can actually be useful to strengthen the relationship between wine and Italian young adults, and also as a marketing tool, it is necessary to analyze in depth the attitude of young Italians towards sustainability attributes of wine.

8.3 METHODOLOGY

For the purposes of our research experimental auctions were adopted due to their ability to induce each bidder to reveal his or her truthful value for the good (Lusk and Shogren, 2007). Participants were recruited among undergraduates in the city of Naples, Italy.[1] The only requirements were to be a wine consumer (at least once a month) and to be aged among 18 and 35. The data were collected between winter 2012 and summer 2013. In all, eight sessions were held, with 10 participants in each session (n=80). Participants were paid €10 for one and a half hours of their time and received an ID number. Each session started with two training auctions (with potato chips and a chocolate snack) where participants were encouraged to ask questions and expose potential uncertainties (see Box 1 below for details).

BOX 1: OVERVIEW OF THE EXPERIMENTAL PROCEDURE.

- Step 1: All students signed a consent form and a form committing them to buy the wine if they won the auction.
- Step 2: Participants were fully briefed on the procedure of the auction method using a PowerPoint presentation and a script where a short example on how bids are sorted in descending order and on how the 5th highest bid and the winner are selected. Also, a numerical example was given to show respondents why it is in their best interest to bid exactly the amount the product is worth to them.
- Step 3: Two training auctions were conducted using potato chips and a chocolate snack.
- Step 4: Participants were handed out the four wine bottles to look closely for differences and information cues.
- Step 5: The fifth-price auction was performed with five rounds.
- Step 6: Participants completed a questionnaire.
- Step 7: Each participant went to the cashier and received €10 for taking part in the auction minus eventual payment for winning.

Respondents were asked to complete a short questionnaire after finishing the auction. Information was collected on socio-demographics, lifestyle, attitudes towards the environment and society, wine consumption habits and knowledge of sustainability practices and specific wine labels. Since no comprehensive sustainability label for wine is available on the Italian market, we asked respondents to bid on four different products: a conventional wine, a wine with a carbon neutral logo showing a green footprint with the writing CO2, a wine including a Libera Terra logo and a wine with the Wine for Life logo (see Box 2 for a detailed description of the specific meanings of the logos).

BOX 2: LABEL EXPLANATION.

- Carbon neutral indicates that all of the greenhouse gases released during wine production, packaging and delivery have been reduced to zero, making this a wine which does not impact negatively on climate change.
- Centopassi - Libera Terra is a label used to commercialize wines produced by Libera Terra Cooperatives that only use land confiscated from the crimi-

nal organization (Mafia) in the Upper Belice Corleonese area (Trapani and Agrigento provinces, Sicily).
- Wine for Life is an initiative of the Community of Sant'Egidio, which through a label on the wine bottle indicates the winery's commitment to social responsibility. Specifically, producers pay half a euro for each label, all proceeds being used to combat AIDS in Africa.

Each bottle of wine (0.75 l) with the sustainable logo carried a brief explanation of its meaning and purpose. All four wines had the same general information: geographical indication (PGI Sicily), vintage (2011), and type (red). No additional information on brand, varietal grapes used, sensory characteristics or actual market price was given to respondents (see Fig. 1). No reference price was given to respondents since previous scholars have demonstrated that provision of reference or field price information influences bid values in experimental auctions (Drichoutis et al., 2008 and Corrigan and Rousu, 2006). The full bidding approach was used (i.e. asking participants to bid on all the products) as several studies agree that subjects tend to value the auctioned products more in the endowment procedure (Lusk et al., 2004; Corrigan and Rousu, 2006 and Gracia et al., 2011). Finally, ordering effect was resolved through randomization. In the training auctions we posted prices to explain the auction mechanism, but during the wine auctions we did not reveal any bidding information.

Based on the second-price Vickrey auction methodology (Vickrey, 1961), an experimental valuation process using a fifth-price auction was developed. The choice of the fifth highest bid makes it possible to increase the number of participants in the transaction, and hence increase the degree of involvement in the auction of those individuals who attribute low values to the products on sale. As noted by Lusk et al. (2004), this type of auction combines the advantages of second-price and random nth-price auctions. Furthermore, Lusk et al. (2007) demonstrated that if the number of participants who could purchase the product is approximately half the session, all bidders would generally be more engaged. We made it clear to the subjects that only one round and one product would be binding, to avoid demand reductions and wealth effects (Shogren et al., 1994). All data were analyzed with STATA statistical software (version 11.0).

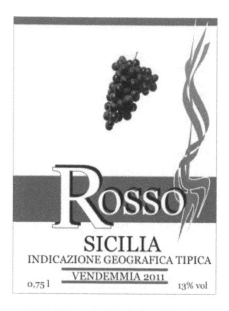

FIGURE 1: Labels used in the experimental auctions.

We are interested in the factors affecting WTP for wines with different attributes. Hence the dependent variable in our model is the average WTP bid for a given subject for each of the products. Given the nature of the data we used a Tobit (Tobin, 1958) model to analyze the bidding behavior for View the MathML source (i=1, 2, 3, and 4) 2. As our interest is in terms of the main effects we ignored possible interactions. In order to determine which estimation method was most appropriate between Tobit and double hurdle, we followed Lusk and Shogren (2007) and calculated a likelihood ratio statistic.

In general, the Tobit model can be expressed as

$$y_i^* = \beta'^{x_i} + u_i, \quad u_i \sim N(0, \sigma^2)$$

$$y_i = y_i^* \text{ if } y_i^* > 0 \text{ or } 0 \text{ if } y_i^* \le 0$$

Therefore the expected willingness to pay for consumer i can be computed as

$$E(y_i^*) = E(y_i|y_i > 0) \times f(y_i|y_i > 0) + E(y_i|y_i = 0) \times F(y_i = 0)$$

$$= [x'\beta + \sigma\lambda(\beta'x|\sigma)] \times \Phi\left(\frac{\beta'xi}{\eth}\right) + 0$$

$$= \beta'^{x_i}\Phi\left(\frac{\beta'xi}{\eth}\right) + \sigma\Phi\left(\frac{\beta'xi}{\eth}\right)$$

where the inverse Mills ratio $\lambda(\beta'x/\sigma)$ is equal to $\varphi(\beta'x/\sigma)/\Phi(\beta'x/\sigma)$ and the marginal effect for the continuous variable x_j is

$$\frac{\delta E(yi)}{\delta xj} = \Phi\left(\frac{\beta' xi}{\eth}\right)\beta_j$$

In particular, the independent variables are participants' socio-demographic and lifestyle characteristics, consumption frequency of wine and other alcoholic beverages, sustainability knowledge and concern. For bid premiums (Carbon neutral bid – conventional bid, Libera Terra bid – conventional bid, and Wine for Life bid – conventional bid) we applied the ordinary least squares method (OLS), as premiums can be positive or negative.

8.4 RESULTS

Due to the specific features of our sample (only undergraduates) several common socio-demographic characteristics were not surveyed since they would not have added useful insights into the respondents' profile (marital status, average annual income, responsibility in everyday food shopping, etc.). Table 1 provides a summary of the independent variable means and standard deviations; 60% of the participants were female, 70% lived in non-urban settings, and the average age was slightly above 23 years. Only 14% of respondents can be considered high consumers of wine (more than twice a week); similarly, consumption per week frequencies for beer, spirits and alcopops were quite low. The main site of wine consumption was away from home (82%). Taking into account participants' concern for sustainability in everyday food shopping, 44% stated they cared. Similar outcomes were found when it came to "caring about environmental sustainability in wine shopping" (49%), while far fewer were concerned about social sustainability in wine shopping (22%). Knowledge of the three labels was quite low: on a scale from 0 to 4 carbon neutral received 1.66 points, followed by Wine for Life with 1.24 and Libera Terra with 1.21.

TABLE 1: Independent variables, included in the estimation models, means and standard deviations (N=80).

Variable Name	Mean	SD	Variable coding
Age	23.3	3.8	Age in years
Gender	0.60	0.47	0=male, otherwise 1
Area of residence	0.71	0.41	0=urban, otherwise 1
Wine consumption frequency per week	1.14	1.07	1=less than twice, 2=2 or 3 times, 3=4 or more times
Wine consumption location (main)	0.18	0.43	0=at home, otherwise 1
Beer consumption frequency per week	1.67	1.92	1=less than once, 2=1 or 2 times, 3=3 or more times
Spirits consumption frequency per week	1.13	0.47	1=less than once, 2=1 or 2 times, 3=3 or more times
Alcopops consumption frequency per week	1.09	0.56	1=less than once, 2=1 or 2 times, 3=3 or more times
Caring about sustainability in everyday food shopping	0.44	0.42	1=very important and important, otherwise 0
Caring about sustainability in everyday non-food shopping	0.83	0.44	1=very important and important, otherwise 0
Caring about environmental sustainability in wine shopping	0.51	0.48	1=very important and important, otherwise 0
Caring about social sustainability in wine shopping	0.22	0.39	1=very important and important, otherwise 0
Knowledge of CN label	1.66	1.24	0–4 (1=low, 4=high)
Knowledge of LT label	1.21	1.10	0–4 (1=low, 4=high)
Knowledge of WFL label	1.24	1.37	0–4 (1=low, 4=high)

As reported in Fig. 2, mean bids for the four wines (considering all five rounds) vary quite widely, as WTP for the conventional wine is €2.50 while WTP for Wine for Life is over 57% higher, reaching €3.93. Carbon-neutral wine WTP is €3.24 whereas Libera Terra wine was valued by respondents at €3.08. All the average differences between conventional wine mean bids and the other three wines are statistically significant according to the Wilcoxon signed-rank test (p<0.001).

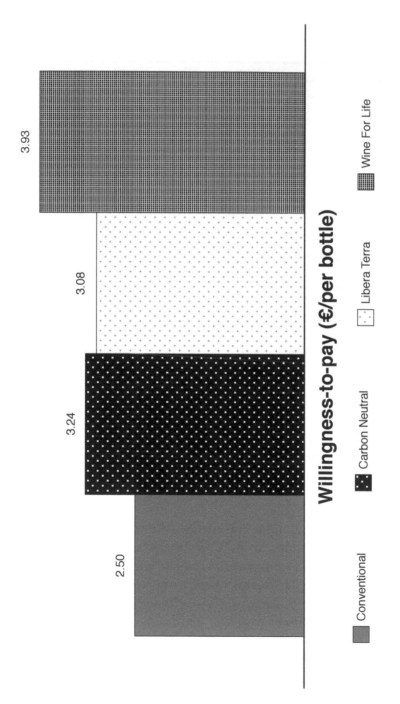

FIGURE 2: Overview of mean bids per auctioned wine.

TABLE 2: Tobit parameter estimates on bids for the four auctioned wines.

Variable	Conventional	Carbon Neutral	Libera Terra	Wine For Life
Constant	1.71	1.43	1.22	1.84
Age	1.85	2.43*	2.22	3.89**
Gender	0.28***	0.36**	0.49**	0.51***
Urban	0.18	0.16	0.12	0.20
Wine consumption frequency	0.21**	0.13**	0.28**	0.71***
Wine consumption location	0.41	0.32	0.09	0.68
Beer consumption frequency	0.03	0.35	0.07	0.61
Spirits consumption frequency	−0.76	−0.65	−0.43	−0.52
Alcopops consumption frequency	−0.04	−0.61	−0.08	−0.23
Caring about sustainability in everyday food shopping	0.30	0.66***	0.43***	0.49**
Caring about environmental sustainability in wine shopping	0.74	0.48**	0.60**	0.02**
Caring about social sustainability in wine shopping	0.12	0.90	0.54	0.02
Knowledge of CN label	0.11	0.26**	0.8	0.76
Knowledge of LT label	0.17	0.22	0.18	0.07
Knowledge of WFL label	0.15	0.24	0.19	0.33**
Likelihood-ratio $\chi2$ (14)	202***	209***	218***	206***
Number of observations	400	400	400	400

*Significance level reported in Tobit models p<0.1. **Significance level reported in Tobit models p<0.05. ***Significance level reported in Tobit models p<0.01.*

Table 2 reports parameter estimates of bid regressions for the four wines considered using all data, i.e. since there are 80 respondents and each of them bid five times, these regressions are based on 400 observations. For the Carbon Neutral wine, the Tobit results in the second column show that six out of 14 estimated parameters are statistically significant: age, gender, wine consumption frequency, caring about sustainability in everyday food shopping, caring about environmental sustainability in wine shopping and knowledge of the specific label. These same variables also explain WTP for Wine for Life and Libera Terra (with the notable exception of knowledge of the label). It is also important to point out that gender and wine consumption frequency are variables also affecting WTP for conventional wine.

TABLE 3: OLS regression results on premiums.

Variable	Premium (CN-Con)	Premium (LT-Con)	Premium (WFL-Con)
Constant	0.11	0.18	0.27
Age	0.21**	0.12**	0.28**
Gender	0.27***	0.13***	0.31***
Urban	0.05	0.16	0.11
Wine consumption frequency	0.19**	0.18*	0.23**
Wine consumption location	−0.12	−0.17	−0.01
Beer consumption frequency	−0.05	−0.06	−0.13
Spirits consumption frequency	−0.08	−0.02	−0.07
Alcopops consumption frequency	0.12	0.09	0.08
Caring about sustainability in everyday food shopping	0.18*	0.07*	0.11*
Caring about environmental sustainability in wine shopping	0.44**	0.34	0.42**
Caring about social sustainability in wine shopping	0.09	0.03	0.11
Knowledge of CN label	0.27**	0.08	0.16
Knowledge of LT label	0.06	0.03	0.01
Knowledge of WFL label	0.07	0.19	0.12*
Log Likelihood	−1104.95	−1292.63	−1585.62
N	400	400	400

*Significance level reported in OLS models $p<0.1$. **Significance level reported in OLS models $p<0.05$. *** Significance level reported in OLS models $p<0.01$.*

To further explore respondents' attitudes toward the sustainable wines we applied OLS regression to understand factors underlying the price premium assigned to these products (ΔWTP sustainable wine – conventional wine). As shown in Table 3 the significant variables explaining all price premiums for all products are: age, gender, wine consumption frequency and caring about sustainability in everyday food shopping. Interestingly, caring about wine sustainability is statistically significant for the carbon neutral wine and wine for life premiums, but not for the Libera Terra wine. Similarly, knowledge of the specific label impacts price premiums only for carbon neutral and Wine for Life. In particular, the estimated coefficients for age show that older participants tend to bid higher for the three sustainable wines; similarly, females reveal higher WTPs for these wines. No other variables considered appear to have a significant influence on WTP premiums.

8.5 DISCUSSION AND CONCLUSION

As noted by Schmit et al. (2013), at present the wine industry's sentiment is that consumers consider organic wine an inferior product while eco-certifications might grant broader benefits that go beyond price premium.3 Indeed, this idea would appear to be substantiated by several studies: Loureiro (2003) estimated that Colorado environmentally friendly wines receive a small premium compared to conventional wines. Similarly, Bazoche and colleagues (2008) proved that wines with environmental characteristics do not seem to be valued more highly than traditional Bordeaux. Moreover, Delmas and Grant (2010) showed that eco-labeling has a negative impact on prices for organic California wines, while there is a price premium associated with eco-certification. Furthermore, previous papers have also revealed that consumers' level of environmental knowledge influences their willingness to purchase more environmentally friendly wines (Barber et al., 2009); whereas other scholars (Brugarolas et al., 2005) show that consumers with healthier lifestyles tend to pay higher prices for organic wines. Recent findings of Mueller and Remaud (2013) reveal that marginal WTP for environmentally responsible claims is about three times as high as for the specific socially responsible claim;and while the WTP for environmental responsibility is non-negative across all the investigated markets, it is negative for the socially responsible claim in France and Francophone Canada.

Alongside the above-portrayed market scenario, wineries in the New World seem to be currently more sensitive to environmental and social issues connected to wine production processes (Cholette et al., 2005). In addition, the literature is somewhat conflicting on the importance for Millennials of environmental and social issues in the wine sector (MacDonald et al., 2013 and Thach and Olsen, 2006).

Our results indicate that consumers value sustainability attributes of wine positively. This result is robust across all the products auctioned, as WTPs for all three sustainable wines were significantly higher. The average premium that young individuals were willing to pay for a sustainable wine ranges between 23% and 57% of the average price of the conventional wine, depending on what feature is considered (social, environmental or solidarity). Moreover, age, gender, wine consumption frequency and caring about sustainability in everyday food shopping significantly affect these premiums. Particularly notable is the outcome that female and older respondents tend to bid higher for the three wines considered sustainable.

Our findings should prove particularly useful for marketers and entrepreneurs since studies that compare different sustainable aspects of wine are particularly scant and no research has so far focused on young individuals. In addition, the young adult cohort is an attractive segment for multinational firms across the globe, particularly in emerging markets (Douglas and Craig, 2006, Kjeldgaard and Askegaard, 2006 and Thach and Olsen, 2006). This article also contributes to the growing literature on consumer valuation of sustainable labels for foods (e.g. McCluskey et al., 2009, Annunziata et al., 2011 and Vecchio and Annunziata, 2013).

In addition, this research provides a number of insights into the characteristics of young wine consumers in Italy and, to the extent that these findings apply more generally, it contributes to a very limited European literature. However, there are several limitations inherent in this type of study, a few of which are worth mentioning. First and foremost there are strong social desirability issues (Fisher and Katz, 2000), as respondents often seek to satisfy social norms rather than reveal their true preferences. Furthermore, the decision to include in the research only three wines with particular social, ethical or environmental features may have an impact on overall findings (the addition of other certifications may influence specific WTPs, such as the organic label). The final number of subjects involved in the experimental treatments was quite limited (n=80). Though this is an acceptable sample size in the literature, it would lend more credibility if we had a larger sample. Additionally, for our convenience we recruited only undergraduate students, while involving young adult consumers in general (older and responsible for household food shopping) and in a real market environment (i.e. supermarket or wine store) could have ensured a stronger representativeness of actual wine shoppers (even if in the literature there are contrasting opinions, (see Chang et al., 2009; Depositario et al., 2009)). Moreover, the specific characteristics of the auction protocol, the Vickrey methodology, the absence of price references and non-earned rewards (e.g. Vecchio and Pomarici, 2013), undoubtedly influenced respondents.

Overall, there are several straightforward extensions of the current work. For example, our focus was on the young adult cohort, but further research on the entire Italian population of wine shoppers would provide useful insights about differences in WTP behavior and attitudes toward sustainability issues. Furthermore, a comparison with young wine drinkers of dissimilar countries with different consumption habits and food traditions would also be of interest and could yield divergent findings. Future research should try to replicate our experiment using other mechanisms to test the robustness of our findings, particularly when other cues are included in the valuation scenario (such as peers' opinions, shopping environment, public campaigns, and third-party certifications). Finally, integrating sensory evaluation of the products in this type of experiment appears particularly important since previous research demonstrated that quality aspects and sensory evaluation dominate all other extrinsic environmental factors (Schmit et al., 2013).

REFERENCES

1. Anderson, K., Nelgen, S., 2011. Wine's Globalization: New Opportunities, New Challenges. Working Paper No. 0111, Wine Economics Research Centre, University of Adelaide, June.
2. A. Annunziata, P. Pascale, S. Ianuario Consumers' attitudes toward labelling of ethical products: The case of organic and fair trade products J. Food Prod. Mark., 17 (5) (2011), pp. 518–535
3. P. Auger, T.M. Devinney Do what consumers say matter? The misalignment of preferences with unconstrained ethical intentions J. Bus. Eth., 76 (4) (2007), pp. 361–383
4. N. Barber, D.C. Taylor, C.S. Deale Wine tourism, environmental concerns, and purchase intention J. Travel Tour. Mark., 27 (2) (2010), pp. 146–165

5. N. Barber, C. Taylor, S. Strick Wine consumers' environmental knowledge and attitudes: influence on willingness to purchase Int. J. Wine Res., 1 (2009), pp. 59–72

6. Bazoche, P., Deola, C., Soler, L.G., 2008. An Experimental Study of Wine Consumers' Willingness to Pay for Environmental Characteristics. Selected Paper, 12th Congress of the European Association of Agricultural Economists, Ghent, Belgium.

7. M. Brugarolas Molla-Bauza, L. Martinez-Carrasco Martinez, A. Martinez Pveda, M. Rico Perez Determination of the surplus that consumers are willing to pay for an organic wine Span. J. Agric. Res., 3 (2005), pp. 43–51

8. M. Carrington, B. Neville, G. Whitwell Why Ethical consumers don't walk their talk: towards a framework for understanding the gap between the ethical purchase intentions and actual buying behaviour of ethically minded consumers J. Bus. Eth., 97 (1) (2010), pp. 139–158

9. J. Chang, J. Lusk, F. Norwood How closely do hypothetical surveys and laboratory experiments predict field behavior? Am. J. Agric. Econ., 91 (1) (2009), pp. 518–534

10. K.L. Christ, R.L. Burritt Critical environmental concerns in wine production: an integrative review J. Clean. Prod., 53 (2013), pp. 232–242

11. Cholette, Susan, Richard Castaldi, April Frederick, 2005. Globalization of the Wine Industry: Implications for Old and New World Producers. International Business and Economy Conference Proceedings. January.

12. P. Combris, P. Bazoche, E. Giraud-Héraud, S. Issanchou Food choices: What do we learn from combining sensory and economic experiments? Food Qual. Preference, 20 (8) (2009), pp. 550–557

13. J.R. Corrigan, M.C. Rousu Posted prices and bid affiliation: Evidence from experimental auctions Am. J. Agric. Econ., 88 (4) (2006), pp. 1078–1090

14. Delmas, M.A., Grant, L.E., 2010. Eco-labeling strategies and price-premium: the wine industry puzzle. Bus. Soc. (published online 11 March 2010).

15. D. Depositario, R. Nayga Jr., X. Wu, T. Laude Should students be used as subjects in experimental auctions? Econ. Lett., 102 (2009), pp. 122–124

16. S.P. Douglas, C.S. Craig On improving the conceptual foundations of international marketing J. Int. Mark., 14 (1) (2006), pp. 1–22

17. A. Drichoutis, P. Lazaridis, R.M. Nayga The role of reference prices in experimental auctions Econ. Lett., 99 (3) (2008), pp. 446–448

18. R.J. Fisher, J.E. Katz Social-desirability bias and the validity of self reported values Psychol. Mark., 17 (2000), pp. 105–120

19. S.L. Forbes, D.A. Cohen, R. Cullen, S.D. Wratten, J. Fountain Consumer attitudes regarding environmentally sustainable wine: an exploratory study of the New Zealand marketplace J. Clean. Prod., 17 (13) (2009), pp. 1195–1199

20. B. Gabzdylova, J.F. Raffensperger, P. Castka Sustainability in the New Zealand wine industry: drivers, stakeholders and practices J. Clean. Prod., 17 (2009), pp. 992–998

21. A. Gracia, M. Loureiro, R.M. Nayga Valuing an EU animal welfare label using experimental auctions Agric. Econ., 42 (2011), pp. 669–677

22. ISPO-Federvini (2012). I Giovani e la Cultura del Bere.

23. D. Kjeldgaard, S. Askegaard The glocalization of youth culture: the global youth segment as structures of common difference J. Consum. Res., 33 (2006), pp. 231–247

24. C. Lange, C. Martin, C. Chabanet, P. Combris, S. Issanchou Impact of the information provided to consumers on their willingness to pay for Champagne: comparison with hedonic scores Food Qual. Preference, 13 (7) (2002), pp. 597–608

25. L. Lockshin, A.M. Corsi Consumer behaviour for wine 2.0: a review since 2003 and future directions Wine Econ. Policy, 1 (2012), pp. 2–23

26. M.L. Loureiro Rethinking new wines: Implications of local and environmentally friendly labels Food Policy, 28 (2003), pp. 547–560

27. J. Lusk, T. Feldkamp, T. Schroeder. Experimental auction procedure: impact on valuation of quality differentiated goods Am. J. Agric. Econ., 86 (2004), pp. 389–405

28. J.L. Lusk, C. Alexander, M.C. Rousu Designing experimental auctions for marketing research: the effect of values, distributions, and mechanisms on incentives for truthful bidding Rev. Mark. Sci., 5 (3) (2007) (Article 3)

29. J.L. Lusk, J.F. Shogren Experimental Auctions: Methods and Applications in Economic and Marketing Research Cambridge University Press, Cambridge, UK (2007)

30. J.B. MacDonald, A.J. Saliba, J. Bruwer Wine choice and drivers of consumption explored in relation to generational cohorts and methodology J. Retailing Consum. Serv. (2013), pp. 349–357

31. A. Mariani, F. Napoletano, E. Pomarici Small wine-importing countries: dynamic and competitive performance of suppliers Le Bull. de l'OIV, 84 (2011), pp. 968–970

32. A. Mariani, E. Pomarici, V. Boatto The international wine trade: recent trends and critical issues Wine Econ. Policy, 1 (2012), pp. 24–40

33. J.J. McCluskey, C.A. Durham, B.P. Horn Consumer preferences for socially responsible production attributes across food products Agric. Resour. Econ. Rev., 38 (2009), pp. 345–356

34. S. Mueller, H. Remaud Impact of corporate social responsibility claims on consumer food choice: a cross-cultural comparison Br. Food J., 115 (1) (2013), pp. 142–166

35. M.E. Pullman, M.J. Maloni, J. Dillard Sustainability practices in food supply chains: how is wine different? J. Wine Res., 21 (1) (2010), pp. 35–56

36. Rabobank, 2003. Wine is Business: Shifting Demand and Distribution: Major Drivers Reshaping the Wine Industry: F&A Review. Rabobank International. Sáenz-Navajas et al., 2013

37. M.P. Sáenz-Navajas, E. Campo, A. Sutan, J. Ballester, D. Valentin Perception of wine quality according to extrinsic cues: The case of Burgundy wine consumers Food Qual. Preference, 27 (1) (2013), pp. 44–53

38. T.M. Schmit, B.J. Rickard, J. Taber Consumer valuation of environmentally friendly production practices in wines, considering asymmetric information and sensory effects J. Agric. Econ., 64 (2) (2013), pp. 483–504

39. J.F. Shogren, S.Y. Shin, D.J. Hayes, J.B. Kliebenstein Resolving differences in willingness to pay and willingness to accept Am. Econ. Rev., 84 (1994), pp. 255–270

40. L. Thach, J.E. Olsen Market segment analysis to target young adult wine drinkers Agribusiness, 22 (3) (2006), pp. 307–322

41. J. Tobin Estimation of relationships for limited dependent variables Econometrica, 26 (1) (1958), pp. 24–36

42. UNEP, 2011. Visions for Change – Recommendations for Effective Policies on Sustainable Lifestyles.

43. R. Vecchio, A. Annunziata Consumer attitudes to sustainable food: a cluster analysis of Italian university students New MeditMediterr. J. Econ., Agric. Environ., 12 (2) (2013), pp. 47–55

44. R. Vecchio, E. Pomarici An empirical investigation of rewards' effect on experimental auctions outcomes Appl. Econ. Lett., 20 (14) (2013), pp. 1298–1300

45. I. Vermeir, W. Verbeke Sustainable food consumption: exploring the consumer attitude-behaviour intention gap J. Agric. Environ. Eth., 19 (2006), pp. 169–194

46. W. Vickrey Counter speculation, auctions, and competitive sealed tenders J. Finance, 16 (1961), pp. 8–37

CHAPTER 9

SUSTAINABLE CERTIFICATION FOR FUTURE GENERATIONS: THE CASE OF FAMILY BUSINESS

MAGALI A. DELMAS AND OLIVIER GERGAUD

9.1 INTRODUCTION

> We do not inherit the earth from our parents; we borrow it from our children.
> —Antoine de Saint-Exupéry, *Terre des Homes* 1939

Business sustainability has been defined as meeting current needs without compromising the ability of future generations to meet their own needs (World Commission on Environment and Development, 1987). Researchers have argued that the current economic paradigm is not conducive to business sustainability because it places more value on short-term profit

Sustainable Certification for Future Generations: The Case of Family Business. Family Business Review *2014, Vol. 27(3) 228–243 © The Author(s) 2014. Used with permission from the authors.*

motivations than on the longer-term impacts on society, the environment and future generations (Gladwin, Kennelly, & Krause, 1995). Some scholars have called for a modified paradigm that would reconcile short- and long-term orientations and align social, environmental, and economic goals (Gladwin et al., 1995; Slawinski & Bansal, 2009). In this article, we propose a framework that includes future generations as an important stakeholder, driving the adoption of sustainable practices. In the context of family-owned businesses, we develop a perspective in which anticipation of the needs of future generations via the owner's intention of transgenerational succession encourages business sustainability.

A broad literature has emerged over the past decade demonstrating that firms' environmental strategies and practices are influenced by stakeholders, including nongovernmental organizations and employees (Aragón-Correa, 1998; Delmas, 2001; Delmas & Toffel, 2004; Sharma & Henriques, 2005). However, few articles focus on family enterprises (Berrone, Cruz, Gómez- Mejía, & Larraza-Kintana, 2010; Craig & Dibrell, 2006; Neubaum, Dibrell, & Craig, 2012; Sharma & Sharma, 2011), although family-controlled businesses represent approximately 80% of all business enterprises (Gersick, Davis, Hampton, & Lansberg, 1997; Gomez-Mejia, Larraza-Kintana, & Makri, 2003). Most important, family-controlled businesses have been shown to be particularly effective at embracing demands from their internal and external stakeholders (Neubaum et al., 2012) and demonstrate higher levels of investments in proactive environmental practices than nonfamily businesses (Berrone et al., 2010; Craig & Dibrell, 2006; Sharma, & Sharma, 2011). A better understanding of the characteristics that explain family businesses' superior investment in sustainability practices can enrich the stakeholder literature. The examination of sustainability in family businesses is important because many businesses include family ties and relationships.

Many definitions have been provided of family firms, but an important characteristic is that the business is potentially managed across generations of the same family (Chua, Chrisman, & Sharma, 1999; Sharma, Christman & Chua, 1997). This feature is said to have significant bearing on many decisions of family businesses and on performance, including innovation (De Massis, Frattini, Pizzurno, & Cassia, 2013). However, it is unclear under what circumstances this intergenerational feature is initiated and how it affects business sustainability. While the literature has examined the drivers of an effective succession process (Sharma, Chrisman, & Chua, 2003; Sharma, Chrisman, Pablo, & Chua, 2001), it has not yet addressed the question of how the intention for transgenerational succession influences the adoption of sustainable practices. In the context of family businesses, we argue that future generations have a stake in the long-term performance of a business owned by their family and that business owners who are planning their succession are more likely to recognize the needs of future generations and to adopt sustainable practices. We argue that one key explanation for the adoption of sustainable practices is the long-term economic viability of the business that eco-certification can bring to these future generations. To understand the link between intergenerational intent and business sustainability, we used data from a survey of 281 wineries in the United States, including information on the intention to pass down the winery to family members and information about eco-certification practices.

Business sustainability can take many forms. In this article, we focus on eco-certification, which represents the adoption of codified environmental practices and the certification of these practices by a third party (Delmas & Grant, 2014). As eco-certification is associated with third party verification, it provides researchers with confidence in the adoption of substantive environmental practices, limits concerns of greenwashing (Delmas & Burbano, 2011), and functions as an effective signaling mechanism of the firm's environmental performance (Delmas, 2002; Jiang & Bansal, 2003; King, Lenox, & Terlaak, 2005). Furthermore, eco-certification has been shown to facilitate efficiency gains and improvement in product quality (Rondinelli & Vastag, 2000). Because of the possible link between eco-certification and performance, eco-certification has been portrayed has one of the most promising forms of business sustainability (Delmas & Young, 2009). This important characteristic makes eco-certification particularly suitable to understand the role of economic motivations, such as increasing market share or producing higher quality products, in business sustainability decisions. Furthermore, eco-certification represents various shades of green since some firms can adopt eco-certification for only a few products while other can certify all their products and processes. This allows us to go beyond a dichotomous analysis of business sustainability that contrasts brown firms to green firms and instead focus on the drivers of different levels of commitment to business sustainability. It also allows us to assess whether firms with low commitment toward business sustainability differ significantly from those with no commitment.

Our research contributes to both the stakeholder and the family business enterprise literatures. First, we integrate future generations into the stakeholder perspective. The stakeholder literature focuses mostly on current stakeholders, and while successors have been described by Sharma et al. (2001) as an important stakeholder group with a legitimate claim on the firm and a legitimate concern over the succession process, we still have little understanding of how the prospect of succession affects decisions to adopt sustainable practices by current owners. Second, we are able to identify business sustainability as an additional outcome of transgenerational intention to those previously studied in the family business literature. Third, because eco-certification potentially leads to increased quality of products and soil, and improved signaling about business sustainability to stakeholders, we ascertain the role of quality and market considerations as important motivators to adopt sustainable practices in addition to the desire to maintain socioeconomical wealth previously identified in the family business literature.

9.2 FUTURE GENERATIONS AS A STAKEHOLDER

Freeman (1984) defines a stakeholder as "a group or individual who can affect or is affected by the achievement of the organization's objectives" (p. 46). The stakeholder approach proposes that firms should not only focus their strategic decision on generating shareholder value, but should also include the interests of a variety of stakeholders such as employees, customers, communities, the media, and regulatory agencies (Delmas & Toffel, 2004). The explanatory power of stakeholder analyses has been shown in a variety of research in the environmental management literature (Buysse & Verbeke, 2003; Delmas & Toffel, 2008; Henriques & Sadorsky, 1996; Sharma & Henriques, 2005) and in family business (Bingham, Dyer, Smith, & Adams, 2011; Neubaum et al., 2012). In particular, Neubaum et al. (2012) have shown that attention to family employees along with concern for environmental protection help family firms' performance.

Future generations can be thought as a stakeholder that is particularly salient for family firms as compared with nonfamily firms (Bingham et al., 2011; Sharma et al., 2001). Indeed, family business enterprises differ in many dimensions from other businesses. One of these dimensions consists of the handling of succession, which refers to all activities related to the transition of the business from one generation to the next (Barry, 1975; Sharma et al., 2001), and that often remains in the family. The succession process is defined as "the actions and events that lead to the transition of leadership from one family member to another in family firms" (Sharma et al., 2001, p. 21). Intergenerational succession can only occur if there is a family member willing to take over the leadership. Research has therefore suggested future generations as potential stakeholders in the succession process, since they affect or can be affected by leadership transitions (Sharma et al., 2001).

Future generations possess several elements that qualify them as a stakeholder. First, several scholars define stakeholders in terms of their necessity for the firm's survival (Bowie, 1988; Freeman & Reed, 1983). Heirs are necessary for the survival of the business as a family business. Mitchel, Agle, and Wood (1997) differentiated further between groups that have a legal, moral, or presumed claim on the firm and groups that have an ability to influence the firm's behavior, direction, process, or outcomes. Heirs are part of both of these groups, since they have a presumed claim on the family business because of their lineage and have the ability to influence the firm's behavior once they inherit the family firm. Scholars have further differentiated between current and potential stakeholders. For example, Starik (1994) refers to stakeholders as those who "are or might be influenced by, or are or potentially are influencers of, some organization" (p. 90). As intergenerational succession is a future event, heirs can therefore be a subset of potential stakeholders. The fact that heirs will likely inherit a business can influence how the current owner behaves in anticipation of intergenerational succession. Finally, scholars have argued that the concept of stakeholder encompasses a socioemotional dimension, in which stakeholders are partners whose futures and stakes are intertwined (Freeman & Gilbert, 1988; Starik, 1995). This socioemotional dimension is at the core of the relationship between business owners and their heirs and is further evidence of the future generations as stakeholders (Berrone, Cruz, & Gomez-Mejia, 2012).

9.3 HYPOTHESES

9.3.1 INTERGENERATIONAL TIES

Recent research indicates that family businesses tend to show higher levels of corporate social responsibility than other firms (Berrone et al., 2010; Dyer & Whetten, 2006; Post, 1993). These higher levels of investments have been explained by the ability of family business owners to have a longer-term view of their investments. Indeed, owners of family businesses are said to care about the long-term objectives of other family members, and their involvement in the business, more than business owners who do not have family involved in the business and who are said to embrace objectives of a shorter-term nature (Miller, Le Breton-Miller, & Scholnick, 2008).

Two main characteristics of family businesses can facilitate this long-term view. The first relates to the ability of family business owners to make independent decisions. Indeed, family firms, in which ownership and control are often embodied into a single decision maker, may produce different managerial rules for investment decisions than firms in which the ownership and control functions are separated (Fama & Jensen, 1985). Because family firms are owned and managed by family members, they are more able to make unilateral decisions than nonfamily firms where ownership is more dispersed (Carney, 2005). Furthermore, family businesses that are privately owned do not face short-sighted investors who could hamper a longer-term perspective.

The second and, as we argue, most important characteristic lies in the connection of family businesses to the next generation. One important element that distinguishes family businesses from other businesses is "the intention to shape and pursue the vision of the business . . . in a manner that is potentially sustainable across generations of the family or families" (Chua et al., 1999, p. 25). The concept of sustainability across generations indicates intergenerational ties and, therefore, the availability of a family successor (Chua et al., 1999). The long-term perspective of family is related to membership in a family system: owners of family businesses invest in building the business for the long-run benefit of various family members (Gomez-Mejia, Takacs- Haynes, Nuñez-Nickel, Jacobson, & Moyano-Fuentes, 2007; Habbershon & Pistrui, 2002; James, 2006). The extension of horizons to the next generation "acts as an incentive for proprietors to postpone consumption out of a concern for the welfare of the proprietors' children, grandchildren, as well as other family members" (James, 1999, p. 47). Eco-certification, by reducing the environmental footprint of the business, allows family businesses owners to invest in the long-term sustainability of their business for the benefit of the next generation.

Factors hampering this long-term perspective have been shown as the inability of the owner to pass the business on to their children or other problems that interfere with the overlapping generational features of the family firm (James, 1999). These can include conflict with inheritance plans, lack of heirs in the family, or open unwillingness of heirs to take over the family business. Such factors reduce the incentives for the family business owner to make investments beyond his or her expected life.

Here, we contend that the ability of the owner of family firms to extend the horizon beyond their expected lives is activated when the owner intends to pass down its business to his or her heirs. In the case of eco-certification, we argue that owners who intend to pass down their business to future generations are more likely to adopt eco-certification for their products than family business owners who do not have this intention. Indeed, intergenerational succession is not an automatic process and requires important preconditions such as the willingness of the incumbent to step aside, the presence of a family successor, and trust in the successor's ability and intentions (Sharma et al., 2001). It is in anticipation of intergenerational succession that the business owner can be influenced by future generations. We therefore hypothesize the following:

Hypothesis 1: Family business owners' transgenerational succession intention is positively associated with the adoption of eco-certification.

9.3.2 INTERGENERATIONAL TIES AND QUALITY MOTIVATIONS

The long-term view of business performance with intergenerational intention might be heightened if the adoption of sustainable practices can increase the long-term quality of products and solidify the business for future generations by improving the winery brand. This is particularly true in agriculture where ecologically sound management can improve soil quality and productivity (Organisation for Economic Cooperation and Development, 2011). The use of organic grown grapes has also been shown to result in superior wine quality because organic growing leads to optimum expression of the land in wine (Delmas & Grant, 2014).

However, this potential increase in quality associated with eco-certification is longer term. Indeed, eco-certification can be a complex and difficult lengthy process. For example, it takes at least 3 years to obtain organic certification, and during that time, the family business owner cannot benefit from the potential price premium associated with certification (Delmas, Doctori-Blass, & Shuster, 2008). Therefore, intergenerational intentions will make these longer-term quality motivations more appealing. Family business owners who seek to increase the quality of their crops and the sustainability of their land might be more likely to adopt certification. In the case of family wineries, the owner might use eco-certification to protect the quality of the soil and the products over the long run. Therefore, a long-term perspective is essential to consider the potential increase in the quality of the product. We hypothesize that the quality motivation will be activated for family business owners who intend to pass down their business to their children. In that case, business owners view their children as having an important stake in the future quality of the product and related economic viability of the family business. Thus, we hypothesize the following:

Hypothesis 2: The positive relation between quality motivation and eco-certification will be stronger in firms that intend to pass down their business to their heirs.

9.3.3 INTERGENERATIONAL TIES
AND MARKET MOTIVATIONS

Another explanation for the adoption of eco-certification includes the objective to build market share and more enduring relationships with customers (James, 2006; Miller & Le Breton-Miller, 2003). Research indicates that family business owners are particularly attentive to their stakeholders and seek to build strong connections with outside stakeholders, and particularly with customers who can sustain the business in times of trouble (Berrone et al., 2012; Gomez-Mejia, Nuñez-Nickel, & Gutierrez, 2001). As we argue below, eco-certification can help family businesses create deeper connections with their current customers and help them reach out to new customers.

One of the objectives of eco-certification and their associated labels is to provide credible information related to the environmental attributes of the product and to signal that the product is superior in this regard to a nonlabeled product (Crespi & Marett, 2005). The assumption behind eco-labels is that environmentally responsible consumers can make informed purchasing choices based on product-related environmental information (Leire & Thidell, 2005). Family business owners might, therefore, seek to solidify or expand their relationships with customers through eco-certification. Ecocertification can help firms gain access to emerging green markets and build long-term customer relationships based on sharing sustainable values (Delmas, 2001). Eco-certification can therefore help to create new and stronger connections with customers.

However, there is still some uncertainty of the value of eco-certification. For example, studies have shown that the presence of competing eco-certification systems has led to consumer confusion about the value of eco-certification (Delmas, Nairn-Birch & Balzarova, 2013; Leire & Thidell, 2005). This is particularly salient as it relates to eco-certification in the wine industry (Delmas & Grant, 2014; Delmas & Lessem, in press). In other words, eco-certification might have the potential for market appeal, but there is some uncertainty on when this appeal will be realized. This is why family business owners who intend to pass down their business to their heirs and have a longer-term vision of their business should be more likely to adopt eco-certification for its future market potential.

Such family business owners will open market opportunities for the future generations. For these reasons, we hypothesize the following:

Hypothesis 3: The positive relation between market motivation and eco-certification will be stronger in firms that intend to pass down their business to their heirs.

9.4 METHODOLOGY

We used the wine industry to test our hypotheses. The wine industry is an excellent context to test the drivers of proactive environmental strategy in family firms (Sharma & Sharma, 2011). First, it is composed of both family and nonfamily firms with different succession practices. Second, wine industries in many countries face a wide array of environmental concerns and increasing pressures to improve their environmental performance (Marshall, Akoorie, Hamann, & Sinha, 2010). Third, firms can adopt several eco-certification systems including organic and biodynamic certification (Delmas & Grant, 2014). However, up until now relatively few scholars have investigated proactive environmental strategies in the wine industry context (Cordano, Marshall, & Silverman, 2010; Delmas & Grant, 2014; Marshall, et al., 2010).

9.4.1 DATA COLLECTION

Because there was no existing publicly available data on the subject, the best method to obtain this information was to directly question wineries and vineyards in California through the dissemination of an online survey. California accounts for an estimated 90% of the US wine production, making more than 260 million cases annually, and consists of family-owned wineries and wineries owned by corporations.[1,2] The survey questionnaire included questions about the winery characteristics, such as size and eco-certification status, and motivations to adopt eco-certification.

Population. Our population consisted of all 1,900 California wineries identified in the California Department of Alcoholic Beverage Control database, which includes all wineries legally licensed to sell alcohol within the state. It, therefore, does not include vineyards that produce grapes but no wine. Phone and e-mail contact information was obtained through an Internet search. The survey was addressed to the owner of the winery or the general manager for nonfamily business wineries. We distributed the survey employing several mediums, including, mailing a recruitment letter with the survey link, sending e-mails, and calling wineries and vineyards to ask for their participation in this survey. The survey was kept open for 3 weeks, with two reminder e-mails sent during that period.

Survey Administration. A total of 1,861 letters describing the study with a link to an online survey were successfully delivered. Three emails (delivered over a 2-week period to 1,186 potential respondents) and 849 phone calls were subsequently used to contact wineries to further encourage participation. In total, 378 responses were gathered, reflecting a 20% response rate, which is comparable to other recent research (Chrisman, Chua, Pearson, & Barnett, 2012; Davis, Dibrell, Craig, & Green, 2013; Delmas & Toffel, 2008; Zellweger, Kellermanns, Chrisman, & Chua, 2012; Zellweger, Nason, & Nordqvist, 2012). Out of these responses we retain 281 usable observations.

We tested sample representativeness in several ways. First, we conducted t tests to compare respondents to nonrespondents along three dimensions. We used data on the nonrespondents from the California Department of Alcoholic Beverage Control database. The survey respondents were 7.4% more likely to have obtained eco-certification than the nonrespondents ($p = .01$). However, they did not differ in terms of the number of years in business ($p = .46$). The overrepresentation of eco-certified respondents was to be expected, since such wineries would be more interested in responding to a questionnaire on the motivations for sustainable agriculture. To correct for this bias, we used the sample weight procedure for survey data in Stata and obtained similar regression results as those with the original sample presented in this article.[3] We also tested for nonresponse bias by comparing early and late respondents, since late respondents have been shown to be similar to nonrespondents (Armstrong & Overton, 1977). We created a set of late respondents with those who responded after receiving the third reminder on April 27, 2009 (Cantwell &

Mudambi, 2005). We did not find a significant difference between the late respondents and the other respondents in terms of status (family business vs. nonfamily business: $p = .89$) and eco-certification ($p = .26$).

9.4.2 VARIABLES

Dependent Variable. Our dependent variable represented the percentage of eco-certified production per winery, which was 10.4% on average. Eco-certification represented the adoption of organic certification or biodynamic certification. Of the vineyards, 21% have between 10% and 90% of their products eco-certified, and 2.5% reach 100% of eco-certified products. Conventional wineries are dominant in our sample (76.5%).

Independent Variables. To identify family businesses and family business owners with the intention to pass down their business to the next generation, we included two variables. *Family business* is a dummy variable that represented whether the winery was family-owned as opposed to other forms of private ownership, publicly traded, or part of a cooperative. Family businesses were dominant in our sample (81.9%).[4] *Heir succession* is a dummy variable that identified those producers whose intention was to pass down the business to their heirs. This constituted about one-half (50.2%) of the producers.[5]

Quality motivations and *market motivations* were identified through a factor analysis based on the four following motivation variables: *improved soil quality, improved quality of grapes, increased demand from restaurants and retailers*, and *growing consumer demand*. Motivations were assessed on a 7-point Likert-type scale and the questions were developed based on Delmas and Toffel (2008) and enhanced with input from industry experts. We conducted a factor analysis with Varimax rotation of these variables, which resulted in two factors and explained 82% of the variance. The variables *improved soil quality* and *improved quality of grapes* loaded on the first factor. The variables *increased demand from restaurants and retailers* and *growing consumer demand* loaded on the second factor. The first factor, therefore, represents quality motivations, while the second factor represents market motivations.

Controls. The controls included winery age (6 categories) and size as proxied by the number of cases produced per year (19 categories). Winery age was included because older wineries may be more likely to be at a stage of intergenerational succession. Smaller wineries may also be more likely to be family-owned. Wineries considered here were created 19.83 years ago and sell around 10,250 cases per year on average. Vertical integration was a binary variable for those wineries (83%) that own part or all of the vineyard as compared with purchasing grapes. Vertical integration makes the winery own its vineyard and be more likely to care about long-term soil quality. Last, we controlled for the geographical location of the winery at the county level from a set of four dummy variables for the most represented counties: Napa Valley (22%), Sonoma Valley (27%), San Luis Obispo (7%), Santa Barbara (8%), and others (36%), which was considered as the reference category. This allowed us to control for the level of eco-certification adoption in specific counties that could affect the adoption of eco-certification by a winery.

A Harman's one-factor test was conducted to test for the presence of common method effect. The following variables (heir succession, quality motivations, market motivations, vertical integration, winery age, number of cases produced) were entered into an exploratory factor analysis, using unrotated principal components factor analysis to determine the number of factors that were necessary to account for the variance in the variables.6 In our case, the results of this analysis show that three factors were present, with the first factor explaining only 23% of the variance and the three factors explaining 56% of the total variance. This suggests that common method variance is not of concern and thus is unlikely to confound the interpretation of the results.

The descriptive statistics and the correlation matrix are provided in Table 1.

9.4.3 MODEL

In Model 1, the level of eco-certified production of winery i was seen as a function of *family business*, *quality motivations*, market motivations, and the exogenous controls. In Model 2, we added *heir succession* to the list of regressors of Model 1 to assess the influence of this key dimension on the quality of the fit. In Model 3, we interacted *heir succession* with *quality motivations* and *market motivations* to check whether the impact of heir succession varied with quality and market motivations. This model is used to test Hypotheses 2 and 3.

9.4.4 ESTIMATION STRATEGY

The dependent variable represents the proportion of ecocertified production per winery. It has two important features: it is a rate, and it includes many observations clustered at zero (76.5%) and several observations in the far-right tail of the distribution (2.5% of our sample firms have all their production eco-certified). Our goal here was to model p, the proportion of eco-certified production as a function of a vector of explanatory variables X' with a special emphasis on the role played by heir succession plans in the process. The usual linear regression models assume that data come from a normal distribution with the mean related to its predictors ($Y \sim N(\mu, \varphi)$ and $\mu = X\beta$). But there are obvious occasions when a normal distribution is inappropriate. Proportions fall into this category as they are, by construction, constrained between 0 and 1.

This is the reason why we adopted a generalized linear model (GLM) approach, a flexible generalization of ordinary least squares which is, among others, designed to model how the mean proportion relates to the set of explanatory variables (see Nelder & Wedderburn, 1972). In GLM, each outcome of the dependent variable is assumed to be generated from a particular distribution in the exponential family[8] ($Y \sim P(\mu, \varphi)$), and a link function provides the relationship between the linear predictor and the mean of the distribution function ($g(\mu) = X\beta$).

The expected proportion of eco-certified production, p, may be modeled using a binomial distribution. Papke and Wooldridge (1996) suggest that a GLM with a binomial distribution and a logit link function, which they term the *fractional logit* model, may be appropriate to model such proportion or fraction. Following these authors, popular econometric softwares such as Stata and R use logit as the default—natural/canonical—link.[9] We therefore use the canonical logit link: $g(p) = \ln(p/(1 - p))$.

9.5 RESULTS

9.5.1 GLM REGRESSIONS

In Table 2, we present GLM estimates to test Hypothesis 1 on the effect of *heir succession* on *eco-certification*. Interestingly, Model 1 showed no difference between family businesses and nonfamily businesses, since the coefficient of the variable *family business* is insignificant. The results of Model 2 showed, on the contrary, that *heir succession* has a strong positive and significant (1% level) influence on the percentage of eco-certified wine. The marginal effect is at about 8.9%. These results, therefore, confirm Hypothesis 1.

The coefficient of the variable *quality motivations* was significant at the 1% level (+4.3% per standard deviation, once we controlled for the influence of heir succession in the model). The variable *market motivations* also came out highly significant at the 1% level and of comparable magnitude (+4% per standard deviation). The negative sign for *number of cases produced* indicated that small businesses were more likely to invest in the certification process than bigger ones. The impact of *winery age* was significant and positive and informed us that older wineries were more willing to invest in the green process than younger ones. This is consistent with the literature predicting the effect of local roots on environmental performance to be stronger under family business ownership (Berrone et al., 2010). *Vertical integration* had no significant influence on ecocertification, which might be explained by the fact that most of our wineries were vertically integrated.

To test Hypotheses 2 and 3, we needed to show how quality and market motivations moderate the effect of intergenerational succession on eco-certification. To do so in Model 3 we use heir succession both as a binary

Table 1. Descriptive Statistics.

	Obs.	Mean	SD	Min	Max	ECP	HS	FB	QM	MM	VI	Age	Cases	SV	NV	SB	SLO
Percentage of eco-certified production (ECP)	281	0.104	0.246	0	1	—											
Heir succession (HS)	281	0.502	0.501	0	1	0.22**	—										
Family business (FB)	281	0.819	0.386	0	1	0.08	0.47**	—									
Quality motivations (QM)	281	0.012	0.982	−3.74	1.301	0.19**	0.11⁺	0.07	—								
Market motivations (MM)	281	0.001	1.008	−2.461	1.93	0.16**	0.01	−0.03	−0.02	—							
Vertical Integration (VI)	281	0.826	0.38	0	1	0.10⁺	0.25**	0.10⁺	0.12*	−0.02	—						
Winery age (Age)	281	19.838	5.88	2.5	100	0.15**	0.08	−0.04	0.09	0.02	0.23**	—					
Cases produced (Cases)	281	10,250	1,365	50.5	40,000,000	0.01	0.01	−0.19**	0.02	0.19**	0.03	0.50**	—				
Sonoma Valley (SV)	281	0.27	0.445	0	1	−0.08	−0.01	−0.02	−0.08	−0.01	−0.03	0.03	0.02	—			
Napa Valley (NV)	281	0.217	0.413	0	1	−0.02	0.01	−0.00	0.08	−0.00	0.11⁺	0.30**	0.16**	−0.32**	—		
Santa Barbara (SB)	281	0.085	0.28	0	1	0.02	0.07	0.02	−0.10⁺	−0.04	−0.14*	−0.08	−0.07	−0.14*	−0.16**	—	
San Luis Obispo (SLO)	281	0.068	0.252	0	1	0.13*	0.08	0.04	0.08	0.03	0.07	−0.06	−0.02	−0.16**	−0.19**	−0.08	—

Note. p Values in parentheses.
**p < .01. *p < .05. ⁺p < .1.

Table 2. The Motivations of Eco-Certification (GLM).

	(1)	Marginal effects	(2)	Marginal effects	(3)	Marginal effects
Family Business	0.176 (0.78)	0.176 (0.78)	-0.277 (-1.02)	-0.042 (-0.90)	-0.250 (-0.95)	-0.037 (-0.85)
Heir succession			0.654** (3.15)	0.089** (3.14)	0.602** (2.92)	0.080** (2.86)
Quality motivations (factor)	0.343** (3.64)	0.049** (3.48)	0.326** (3.57)	0.043** (3.45)	0.058 (0.39)	0.008 (0.39)
Market motivations (factor)	0.291** (3.05)	0.041** (3.12)	0.300** (3.22)	0.040** (3.21)	0.430** (2.86)	0.056** (2.79)
Quality motivations × Heir succession					0.404* (2.10)	0.053* (2.08)
Market motivations × Heir succession					-0.156 (-0.84)	-0.020 (-0.83)
Exogenous controls						
Vertical integration	0.082 (0.39)	0.011 (0.41)	-0.094 (-0.41)	-0.013 (-0.40)	-0.067 (-0.30)	-0.009 (-0.29)
Winery age	0.270** (3.15)	0.038** (3.17)	0.273** (3.18)	0.036** (3.27)	0.275** (3.14)	0.036*** (3.31)
Number of cases produced	-0.046* (-1.96)	-0.007* (-1.90)	-0.056* (-2.38)	-0.007* (-2.30)	-0.058* (-2.43)	-0.008* (-2.37)
Wine regions (ref. category: other)						
Sonoma Valley	-0.315 (-1.35)	-0.039 (-1.50)	-0.304 (-1.28)	-0.036 (-1.42)	-0.302 (-1.29)	-0.035 (-1.43)
Napa Valley	-0.270 (-1.31)	-0.035 (-1.40)	-0.279 (-1.38)	-0.034 (-1.46)	-0.282 (-1.38)	-0.034 (-1.48)
Santa Barbara	0.184 (0.52)	0.029 (0.47)	0.031 (0.09)	0.004 (0.08)	0.026 (0.07)	0.004 (0.07)
San Luis Obispo	0.366 (1.34)	0.064 (1.13)	0.305 (1.12)	0.049 (0.97)	0.267 (0.96)	0.041 (0.84)
Constant	-2.130** (-5.10)		-1.926** (-4.80)		-1.969** (-4.91)	
n	281		281		281	
Log pseudo-likelihood	-73.09		-70.41		-69.52	

Note. Robust z-statistics in parentheses; GLM estimates are derived using a canonical logit link and a binomial distribution.
**$p < .01$. *$p < .05$. $p < .1$.

variable (direct effect) but also as an interaction term with both factors. The results from this model are consistent with previous results overall. The direct impact of heir succession is estimated at around 8%. The direct impact of market motivations is at around 5.6% per standard deviation. However, we did not detect any systematic impact of quality motivations (coefficient of quality motivations nonsignificant). This means that wineries with no intergenerational succession planning adopted eco-certification for market motivations only (+4.4% per standard deviation). Wineries with intergenerational succession planning adopted eco-certification both for market motivations (+5.6% per standard deviaton) and quality motivations (+5.3% per standard deviation). These results confirmed Hypotheses 2 and 3.

9.5.1 ROBUSTNESS TESTS

We ran several tests to check the robustness of our model. First, we calculated variance inflated factors (VIF) to test for potential multicollinearity issues. The mean of the VIF analysis was 3. All individual VIF were below 5, except for the control variable representing the age of the winery, which was just below the rule-ofthumb cutoff value of ten for multiple regression models (Hair, Anderson, Tatham, & Black, 1995; Kennedy, 1992; Marquardt, 1970; Neter, Wasserman, & Kutner, 1989). We tested the robustness of the model without this control variable and obtained similar results as those provided below.[10] This indicates that there is no concern for multicollinearity in our regression models.

Second, we ran a series of Logit models to check whether the drivers of the adoption of green practices in the vineyard were stable or varied with the level of certification. This was done by regressing ten Logits (one for each retained specification) for the probability that a firm i certified production exceeds 0%, 10%, 20%, 30%, 40%, 50%, 60%, 70%, 80%, and 90% respectively. In the set of results presented in Table 3 quality motivations and market motivations are both moderated by heir succession.[11]

For easier interpretation, Figure 1A displays the results for owners who anticipate intergenerational succession (heir succession = yes) and Figure 1B for those who do not anticipate intergenerational succession (heir succession = no). Figure 1A indicates that quality motivations are significant for the adoption of eco-certification until 80%. These are, by far, the strongest drivers of eco-certification among these green wineries. The market motivations are only significant above 60% of ecocertification. This might indicate that winery owners with interegenerational intention only see the value of the market signal of eco-certification at higher levels of certification. In both cases, the motivations become insignificant over 80% certification. These results should be interpreted with caution because the very small percentage of eco-certified wineries at these levels might explain this lack of significance. It is also possible that above these threshold emotions are a stronger driver of certification, although we cannot observe this motivation. For those who are not planning intergenerational succession, Figure 1B shows that market motivations are significant until 40% and then become insignificant. Quality motivations are therefore not significant drivers for these winery owners.

In summary, our results confirmed Hypothesis 1 and indicate that anticipation of transgenerational intention was an important driver of eco-certification adoption. We also confirmed Hypothesis 2, showing that this effect was moderated with quality motivations. Indeed, quality motivations tended to dominate market motivations overall and in the case of owners with transgenerational intention. Regarding Hypothesis 3 on the

Table 3. The Drivers of Eco-Certification and Vertical Integration (Logit Analysis on % of Eco-Certified Production).

	>0%	>10%	>20%	>30%	>40%	>50%	>60%	>70%	>80%	>90%
Heir succession: yes										
Quality motivations	0.086 (1.52)	0.139** (3.69)	0.094** (3.02)	0.082** (2.84)	0.064** (2.59)	0.052** (2.64)	0.047* (2.51)	0.034* (2.26)	0.030* (2.01)	0.008 (0.90)
Market motivations	0.049 (1.22)	0.054' (1.83)	0.032 (1.54)	0.029 (1.59)	0.019 (1.44)	0.015' (1.73)	0.023* (2.17)	0.027** (2.76)	0.022* (2.07)	0.011 (1.06)
Heir succession: no										
Quality motivations	0.005 (0.15)	0.029 (0.97)	0.004 (0.22)	0.005 (0.32)	-0.004 (-0.40)	-0.003 (-0.31)	-0.003 (-0.33)	0.000 (0.04)	0.010 (1.13)	0.000 (0.35)
Market motivations	0.087* (2.31)	0.059* (2.26)	0.045* (2.08)	0.036* (2.04)	0.024' (1.73)	0.020 (1.50)	0.022 (1.64)	0.019 (1.62)	0.016 (1.57)	0.001 (0.46)
Exogenous controls										
Family Business	0.063 (1.09)	0.008 (0.16)	-0.014 (-0.30)	0.004 (0.12)	0.004 (0.16)	-0.003 (-0.14)	-0.004 (-0.16)	-0.008 (-0.36)	—	—
Vertical Integration	-0.125 (-1.46)	-0.036 (-0.51)	0.082** (2.70)	0.052** (1.98)	0.022 (1.21)	0.011 (0.63)	0.009 (0.49)	0.006 (0.30)	—	—
Winery age	0.059* (2.26)	0.053** (2.71)	0.040** (2.94)	0.031** (2.75)	0.027** (3.04)	0.020** (2.73)	0.019** (2.63)	0.016* (2.13)	0.015 (1.43)	0.003 (0.72)
Number of cases produced	0.002 (0.21)	-0.004 (-0.67)	-0.012* (-2.53)	-0.011** (-2.59)	-0.009** (-2.44)	-0.005' (-2.10)	-0.005* (-2.17)	-0.004' (-1.75)	-0.005 (-1.33)	-0.001 (-0.71)
Wine regions										
Sonoma Valley	-0.082 (-1.34)	-0.048 (-0.96)	-0.043 (-1.27)	-0.022 (-0.77)	-0.027 (-1.64)	-0.024' (-1.84)	-0.021 (-1.47)	-0.014 (-1.04)	-0.006 (-0.45)	-0.001 (-0.20)
Napa Valley	-0.046 (-0.78)	-0.013 (-0.26)	-0.022 (-0.68)	-0.032 (-1.25)	-0.030' (-1.82)	-0.024' (-1.82)	-0.019 (-1.46)	-0.018 (-1.38)	-0.009 (-0.67)	-0.004 (-1.44)
Santa Barbara	-0.084 (-1.00)	0.009 (0.11)	0.045 (0.59)	0.047 (0.69)	0.029 (0.66)	0.027 (0.72)	0.014 (0.43)	0.024 (0.61)	0.012 (0.40)	—
San Luis Obispo	0.049 (0.54)	0.110 (1.20)	0.042 (0.71)	0.027 (0.59)	0.020 (0.60)	0.020 (0.69)	0.031 (0.89)	0.024 (0.72)	0.024 (0.60)	0.008 (0.57)
Pseudo R^2	0.076	0.140	0.175	0.194	0.272	0.310	0.279	0.280	0.314	0.393
Log pseudo-likelihood	-142	-112	-92	-79	-63	-55	-54	-48	-33	-18
Observations	281	281	281	281	281	281	281	281	194	184

Note. Z-statistics in parentheses. Coefficients reproduced in this table are marginal effects.
**$p < .01$. *$p < .05$. '$p < .1$.

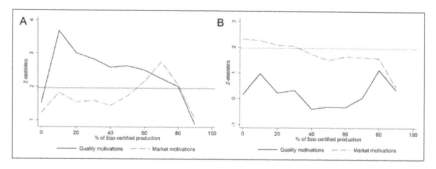

FIGURE 1: Logit regressions—Model 3: (A) Heir succession plan: Yes; (B) Heir succession plan: No.

moderating effect of market motivations, we did find a significant effect of market motivations for family businesses. However, we did not find a significant difference between family business owners who intended to pass down their business to their heirs and the other family firms in terms of market motivations. It seems that market motivations are also present with other family business owners. This could indicate that the current market signal associated with eco-certification is sufficiently strong to appeal to other types of businesses. However, the results from the percentages of adoption showed that market motivation for business without transgenerational intent are only significant for lower levels of adoption (<30%). Such businesses might consider that the market rewards of certification are only rewarding a low commitment.

9.6 DISCUSSION

The literature has described short-term profit motivations as a barrier to the adoption of sustainable practices and has called for the need to develop new management models that include time in the analysis (Slawinski & Bansal, 2009). In this article, we argue that family business owners who intend to pass down their business to their children adopt a longer time frame and are more receptive to the needs of future generations and the sustainability of their business. We show that such businesses are more likely to adopt eco-certification. In doing so, our research contributes to several research perspectives.

While the stakeholder framework has been used to demonstrate how businesses tend to respond to stakeholder pressures by adopting green practices, this literature has mostly ignored family businesses and the connections that businesses make with the future of their own family members. We contributed to the stakeholder literature by showing that future generations should be considered as a main stakeholder, since their existence influences business owners' decisions about eco-certification. Future generations enjoy two main characteristics that qualify them as a stakeholder: They have a presumed claim on the family business because of their lineage, and they have the ability to influence firm behavior once they inherit the family firm. We have described how future generations could influence the adoption of eco-certification, but also impact how current business owners envisage their relationships with their current stakeholders. Our results show that family business owners who intend to pass down their winery to their children are more likely to be responsive to perceived customer demand for green certification. This is consistent with Neubaum et al. (2012), who found a strong relationship between family firm concerns for the environment and concern for their employee well-being. In our case, however, we consider the vision of family involvement in the future of the business as activating the long-term perspective necessary for business owners to embrace business sustainability.

The family business structure and freedom from corporate stakeholders explain why family businesses can make more bold decision because of their independence (Carney, 2005). Here, we find that the private business structure is not enough to explain family business attitude toward business sustainability. Our results confirm that firms that identified themselves as family firms but did not intend to pass down their business to their heirs were not more likely to adopt eco-certification. This lack of intergenerational intent makes such firms comparable with nonfamily private firms. The ability of owners of family firms to extend the horizon beyond their expected lives is only activated when the owners intend to pass down their business to the next generation. These results, in the context of business sustainability, confirm previous literature describing the intergenerational intent as the most important factor differentiating family businesses from other firms (James, 1999). Our results are also consistent with research arguing that the shortcoming of private nonfamily firms, which lack a transgenerational intent, is that they concentrate more on the short-term than family firms do (Miller & Le Breton-Miller, 2005).

While the family business literature has identified the effective drivers of a successful succession planning process (Sharma et al., 2001), it has not yet analyzed how this process could have an impact on the natural environment. We have shown that business sustainability via eco-certification is more likely to be achieved in anticipation of the intergenerational succession process. This finding allows us to isolate one of the specific characteristics of family businesses and to associate it with sustainability. The analysis of the factors that drive family businesses to adopt sustainable practices is important not only because it may be helpful for family-owned firms, but also because many businesses adopt practices that resemble familial ties and relationships.

Traditionally, family businesses have been portrayed as risk averse and conservative (Miller & Le Breton-Miller, 2003). The firm symbolizes the family's heritage and traditions to be maintained over several generations (Berrone et al., 2010). Yet we find family businesses with transgenerational intention to be more innovative than other businesses with the adoption of more advanced sustainable practices. This is consistent with Craig and Dibrell (2006), who found family firms with environmental policies to be more innovative. This raises the question of whether the adoption of such innovative practices is a demonstration of conservative or pioneering behavior. The answer is probably a little of both: a pioneering effort is required to conserve the value of the business. On the one hand, we found that one motivation was to preserve the value of the business for the future generation, and in particular, the quality of the soil and the products. On the other hand, eco-certification is an innovative approach for which the market benefits are still uncertain. It, therefore, requires an investment without immediate return, which is similar to other investments in innovative practices.

Research focusing on family business has emphasized the role of noneconomic factors in the management of family businesses as the key distinguishing feature that separates such firms from other organizational forms (Gomez-Mejia, Cruz, Berrone, & De Castro, 2011). Scholars have argued that because of the ambiguous relationship between the adoption of socially responsible behavior and corporate performance, family firms tend to be more responsive to stakeholders for intangible reasons than for economic reasons (Berrone et al., 2012). Our results complement this perspective, as we show that economic considerations might also play a role in the adoption of eco-certification. Indeed, we find that family businesses with transgenerational intention tend to be motivated by quality objectives that have an impact on long-term economic performance as important drivers for the adoption of eco-certification.

Our findings indicate that anticipation of transgenerational intention is associated with eco-certification adoption and that this effect varies with quality and market motivations. Interestingly, family businesses without intergenerational intention were not motivated by the quality potential associated with eco-certification. This confirms our hypothesis that it is indeed intergenerational intention that activates the will to preserve the quality of the product and the vineyard for the long term. We also find that motivations varied with the level of commitment to eco-certification. Market motivations, for family businesses with intergenerational intention, were a more significant driver than quality considerations for higher levels of certification. For other businesses, market motivations were more significant at lower levels of certification. One explanation for this might be that such businesses considered little market benefit for eco-certification and opted for little commitment. Family businesses with intergenerational intention, because of their longterm perspective, seemed ready for more commitment (i.e., higher levels of certification). This underscores the need to consider certification levels rather than certification as a binary variable, since motivations can vary substantially according to the levels of certification. Research on the adoption of eco-certification has analyzed mostly eco-certification as a binary variable, with adoption and nonadoption being the only alternatives. However, ecocertification rarely covers all the products or activities of the firm, and firms also make decisions on the level of eco-certification they want to adopt. Indeed, firms that have certified 100% of their products are the minority in our sample. We have shown that firms that certify less than 10% of the products have different motivations than those willing to certify the majority or the totality of the products. We also show that wine owners without intergenerational intention are motivated by market motivations for lower levels of certification. It is possible that their shorter-term perspective drives them to adopt ecocertification only for symbolic perspective, to get quick market recognition without substantive commitment to business sustainability. A behavior that could resemble greenwashing by combining positive communication about environmental performance with low environmental performance (Delmas & Burbano, 2011).

Our research is not without limitation. First, our analysis was limited to the California context; future research should explore similar questions in an international setting, as scholars have identified international institutional differences regarding the implementation of environmental practices (Husted, 2005; Husted & Allen, 2006; Darnall, Henriques, & Sadorsky, 2008; Delmas & Montiel, 2008; Delmas & Montes-Sancho, 2011). Second, while our data focused on eco-certification, wineries might adopt other types of sustainable practices that we could not observe through third party certification. Further research should test whether succession intention is also a positive driver of the adoption of these practices. Third, while we asked for information about winery owners' plans for intergenerational intention, we did not ask specific questions about the owner's age or education or the current involvement of family members in the management of the winery. Further research could integrate these additional characteristics. For example, it would be particularly interesting to assess whether family business owners who inherited their winery are more likely to adopt innovative sustainable practices than those who started their own business. Fourth, while our survey included a rich set of variables that allowed us to control for many winery characteristics, its cross-sectional nature hampered us from conducting a dynamic analysis. Further research should examine whether the effects identified in this study persist over time, and should further investigate the precise nature of the dynamic interactions between the firm's external environment (e.g., the existence of informal of formal networks of producers), main business strategy, resources, and organization, and its adoption of eco-certification.

9.7 NOTES

1. U.S. Treasury's Alcohol and Tobacco Tax and Trade division data.
2. USDA, NASS, California field office (2005) California Agriculture Overview.
3. Results available on request.
4. To identify "family firms," we used the following two main questions from the survey questionnaire: (a) "Winery's ownership status?" (Privately owned, Owned by a publicly traded company, Cooperative). If the respondent answered "Yes" on Privately owned, he/she was asked the following second question: (b) "Type of private ownership?" (Family Owned, Company Owned, Partnership with Larger Company). Family owned ownership represents therefore the subset of privately owned wineries that are family owned. 5. To measure "Heir Succession," the question was whether the owner "had family that he/she plans to pass down his/ her winery to."
6. Family business was excluded from the analysis because of his high collinearity with heir succession (i.e., only family business can anticipate to pass down their winery to their heirs). We also excluded the wine region dummy variables.
7. In this theoretical setup, X covers all right-hand-side variables.
8. These include the binomial, gamma, inverse gaussian, negative binomial, poisson, and gaussian distributions.
9. See also Fox (2008), Chapter 15, pp. 382-383.
10. Results available on request.
11. Because of the small number of observations with 60% and above of eco-certified production, achieving convergence with interaction variables between heir succession, quality motivations, and market motivations above this threshold was not possible.

REFERENCES

1. Aragón-Correa, J. A. (1998). Strategic proactivity and firm approach to the natural environment. Academy of Management Journal, 41, 556-567.

2. Armstrong, J., & Overton, T. (1977). Estimating nonresponse bias in mail surveys. Journal of Marketing Research, 14, 396-402.
3. Barry, B. (1975). The development of organization structure in the family firm. Journal of General Management, 2(3), 42-60.
4. Berrone, P., Cruz, C., & Gomez-Mejia, L. R. (2012). Socioemotional wealth in family firms theoretical dimensions, assessment approaches, and agenda for future
5. research. Family Business Review, 25, 258-279.
6. Berrone, P., Cruz, C. C., Gómez-Mejía, L. R., & Larraza-Kintana, M. (2010). Socioemotional wealth and corporate response to institutional pressures: Do family-controlled firms pollute less? Administrative Science Quarterly, 55, 82-114.
7. Bingham, J. B., Dyer, W. G., Jr., Smith, I., & Adams, G. L. (2011). A stakeholder identity orientation approach to corporate social performance in family firms. Journal of Business Ethics, 99, 565-585.
8. Bowie, N. (1988). The moral obligations of multinational corporations. In S. Luper-Foy (Ed.), Problems of international justice (pp. 97-113). Boulder, CO: Westview Press.
9. Buysse, K., & Verbeke, A. (2003). Proactive environmental strategies: A stakeholder management perspective. Strategic Management Journal, 24, 453-470.
10. Cantwell, J., & Mudambi, R. (2005). MNE competence-creatingsubsidiary mandates. Strategic Management Journal, 26, 1109-1128.
11. Carney, M. (2005). Corporate governance and competitive advantage in family-controlled firms. Entrepreneurship Theory and Practice, 29, 249-265.
12. Chrisman, J. J., Chua, J. H., Pearson, A. W., & Barnett, T. (2012). Family involvement, family influence, and family-centered non-economic goals in small firms. Entrepreneurship Theory and Practice, 36, 267-293.
13. Chua, J. H., Chrisman, J. J., & Sharma, P. (1999). Defining family business by behavior. Entrepreneurship Theory and Practice, 23, 19-40.
14. Cordano, M., Marshall, R. S., & Silverman, M. (2010). How do small and medium enterprises go green? A study of environmental management programs in the U.S. wine industry. Journal of Business Ethics, 92, 463-478.
15. Craig, J., & Dibrell, C. (2006). The natural environment, innovation, and firm performance: A comparative study. Family Business Review, 19, 275-288.
16. Crespi, J. M., & Marett, S. (2005). Eco-labeling economics: Is public involvement necessary? In S. Krarup & C. S. Russell (Eds.), Environment, information and consumer behavior (pp. 93-109). Cheltenham, England: Edward Elgar.
17. Darnall, N., Henriques, I., & Sadorsky, P. (2008). Do environmental management systems improve business performance in the international setting? Journal of International Management, 14, 364-376.
18. Davis, W. D., Dibrell, C., Craig, J. B., & Green, J. (2013). The effects of goal orientation and client feedback on the adaptive behaviors of family enterprise Advisors. Family Business Review, 26, 215-234.
19. Delmas, M. (2001). Stakeholders and competitive advantage: The case of ISO 14001. Production and Operation Management, 10, 343-358.
20. Delmas, M. (2002). The diffusion of environmental management standards in Europe and in the United States: An institutional perspective. Policy Sciences, 35(1), 91-119.

21. Delmas, M., & Burbano, V. C. (2011). The drivers of greenwashing. California Management Review, 54(1), 64-87.

22. Delmas, M., Doctori-Blass, V., & Shuster, K. (2008). Ceago Vinegarden. How green is your wine? Environmental differentiation strategy through eco-labels (American Association of Wine Economics Working Paper No. 14). New York, NY: American Association of Wine Economics.

23. Delmas, M. A., & Grant, L. E. (2014). Eco-labeling strategies and price-premium the wine industry puzzle. Business & Society, 53, 6-44.

24. Delmas, M., & Lessem, N. (in press). Eco-Premium or Eco- Penalty? Eco-labels and quality in the organic wine market. Business & Society.

25. Delmas, M. A., & Montiel, I. (2008). The diffusion of voluntary international management standards: Responsible care, ISO 9000 and ISO 14001 in the chemical industry. Policy Studies Journal, 36(1), 65-93.

26. Delmas, M., & Montes-Sancho, M. (2011). US State policies for renewable energy: Context and effectiveness. Energy Policy, 39, 2273-2288.

27. Delmas, M., Nairn-Birch, N., & Balzarova, M. (2013). How managers can choose the most effective eco-labels? Sloan Management Review, Summer.

28. Delmas, M., & Toffel, M. (2004). Stakeholders and environmental management practices: An institutional framework. Business Strategy and the Environment, 13, 209-222.

29. Delmas, M., & Toffel, M. (2008). Organizational responses to environmental demands: Opening the black box. Strategic Management Journal, 29, 1027-1055.

30. Delmas, M. A., & Young, O. R. (2009). Governance for the environment: New perspectives. Cambridge, England: Cambridge University Press.

31. De Massis, A., Frattini, F., Pizzurno, E., & Cassia, L. (2013). Product innovation in family vs. nonfamily firms: An exploratory analysis. Journal of Small Business Management, 51(1). Advance online publication. doi:10.1111/jsbm.12068

32. Dyer, G., & Whetten, D. A. (2006). Family firms and social responsibility: Preliminary evidence from the SP 500. Entrepreneurship Theory and Practice, 30, 785-802.

33. Fama, E. F., & Jensen, M. C. (1985). Organizational forms and investment decisions. Journal of Financial Economics, 14(1), 101-119.

34. Freeman, R. E. (1984). Strategic management: A stakeholder approach. Boston, MA: Pitman.

35. Freeman, R. E., & Gilbert, D. R., Jr. (1988). Corporate strategy and the search for ethics. Englewood Cliffs, NJ: Prentice-Hall.

36. Freeman, R. E., & Reed, D. L. (1983). Stockholders and stakeholders: A new perspective on corporate governance. California Management Review, 25, 93-94.

37. Fox, J. (2008). Applied regression analysis and generalized linear models (2nd ed.). Thousand Oaks, CA: Sage.

38. Gersick, K., Davis, J., Hampton, M., & Lansberg, I. (1997). Generation to generation lifecycles of the family business. Boston, MA: Harvard Business School Press.

39. Gladwin, T. N., Kennelly, J. J., & Krause, T. S. (1995). Shifting paradigms for sustainable development: Implications for management theory and research. Academy of Management Review, 20, 874-907.

40. Gomez-Mejia, L. R., Cruz, C., Berrone, P., & De Castro, J. (2011). The bind that ties: Socioemotional wealth preservation in family firms. Academy of Management Annals, 5, 653-707.

41. Gomez-Mejia, L. R., Larraza-Kintana, M., & Makri, M. (2003). The determinants of executive compensation in family-controlled publicly traded corporations. Academy of Management Journal, 44, 226-237.

42. Gomez-Mejia, L. R., Nuñez-Nickel, M., & Gutierrez, I. (2001). The role of family ties in agency contracts. Academy of Management Journal, 44, 81-95.

43. Gomez-Mejia, L. R., Takacs-Haynes, K., Nuñez-Nickel, M., Jacobson, K. J. L., & Moyano-Fuentes, J. (2007). Socioemotional wealth and business risks in familycontrolled firms: Evidence from Spanish olive oil mills. Administrative Science Quarterly, 52, 106-137.

44. Habbershon, T., & Pistrui, J. (2002). Enterprising families domain: Family-influenced ownership groups in pursuit of transgenerational wealth. Family Business Review, 15, 223-237.

45. Hair, J. F., Jr., Anderson, R. E., Tatham, R. L., & Black, W. C. (1995). Multivariate data analysis (3rd ed.). New York, NY: Macmillan.

46. Henriques, I., & Sadorsky, P. (1996). The determinants of an environmentally responsive firm: An empirical approach. Journal of Environmental Economics and Management, 30, 381-395.

47. Husted, B. W. (2005). Culture and ecology: A cross-national study of the determinants of environmental sustainability. Management International Review, 45, 349-371.

48. Husted, B. W., & Allen, D. B. (2006). Corporate social responsibility in the multinational enterprise: Strategic and institutional approaches. Journal of International Business Studies, 37, 838-849.

49. James, H. (2006). Family capitalism. Cambridge, MA: Belknap–Harvard University Press.

50. James, H. S. (1999). Owner as manager, extended horizons and the family firm. International Journal of the Economics of Business, 6(1), 41-55.

51. Jiang, R. J., & Bansal, P. (2003). Seeing the need for ISO 14001. Journal of Management Studies, 40, 1047-1067.

52. Kennedy, P. (1992). A guide to econometrics. Oxford, England: Blackwell.

53. King, A., Lenox, M., & Terlaak, A. (2005). The strategic use of decentralized institutions: Exploring certification with the ISO 14001 management standard. Academy of Management Journal, 48, 1091-1106.

54. Leire, C., & Thidell, Å. (2005). Product-related environmental information to guide consumer purchases—A review and analysis of research on perceptions, understanding and use among Nordic consumers. Journal of Cleaner Production, 13, 1061-1070.

55. Marshall, R. S., Akoorie, M. E. M., Hamann, R., & Sinha, P. (2010). Environmental practices in the wine industry: An empirical application of the theory of reasoned action and stakeholder theory in the United States and New Zealand. Journal of World Business, 45, 405-414.

56. Marquardt, D. W. (1970). Generalized inverses, ridge regression, biased linear estimation, and nonlinear estimation. Technometrics, 12, 591-612.

57. Miller, D., & Le Breton-Miller, I. (2003). Challenge versus advantage in family business. Strategic Organization, 1(1), 127-134.
58. Miller, D., & Le Breton-Miller, I. (2005). Managing for the long run: Lessons in competitive advantage from great family businesses. Boston, MA: Harvard Business School Press.
59. Miller, D., Le Breton-Miller, I., & Scholnick, B. (2008). Stewardship vs. stagnation: An empirical comparison of small family and nonfamily businesses. Journal of Management Studies, 45, 51-78.
60. Mitchell, R. K., Agle, B. R., & Wood, D. J. (1997). Toward a theory of stakeholder identification and salience: Defining the principle of who and what really counts. Academy of Management Review, 22, 853-886.
61. Nelder, J., & Wedderburn, R. (1972). Generalized linear models. Journal of the Royal Statistical Society, Series A (General), 135, 370-384.
62. Neter, J., Wasserman, W., & Kutner, M. H. (1989). Applied linear regression models. Homewood, IL: Irwin.
63. Neubaum, D. O., Dibrell, C., & Craig, J. B. (2012). Balancing natural environmental concerns of internal and external stakeholders in family and nonfamily businesses. Journal of Family Business Strategy, 3(1), 28-37.
64. Organisation for Economic Cooperation and Development. (2011). A green growth strategy for food and agriculture: Preliminary report. Paris, France: Author. Retrieved from http://www.oecd.org/greengrowth/sustainableagriculture/48224529.pdf
65. Papke, L. E., & Wooldridge, J. (1996). Econometric methods for fractional response variables with an application to 401(k) plan participation rates. Journal of Applied Econometrics, 11, 619-632.
66. Post, J. E. (1993). The greening of the Boston Park Plaza Hotel. Family Business Review, 6, 131-148.
67. Rondinelli, D., & Vastag, G. (2000). Panacea, common sense, or just a label?: The value of ISO 14001 environmental management systems. European Management Journal, 18, 499-510.
68. Sharma, P., Chrisman, J. J., & Chua, J. H. (1997). Strategic management of the family business: Past research and future challenges. Family Business Review, 10, 1-35.
69. Sharma, P., Chrisman, J. J., & Chua, J. (2003). Predictors of satisfaction with the succession process in family firms. Journal of Business Venturing, 18, 667-687.
70. Sharma, P., Chrisman, J. J., Pablo, A. L., & Chua, J. H. (2001). Determinants of initial satisfaction with the succession process in family firms: A conceptual model. Entrepreneurship Theory and Practice, 25(3), 17-36.
71. Sharma, P., & Sharma, S. (2011). Drivers of proactive environmental strategy in family firms. Business Ethics Quarterly, 21, 309-334.
72. Sharma, S., & Henriques, I. (2005). Stakeholder influences on sustainability practices in the Canadian forest products industry. Strategic Management Journal, 26, 159-180.
73. Slawinski, N., & Bansal, P. (2009). Short on time: The role of time in business sustainability. Academy of Management Proceedings (Meeting Abstract Supplement), 1-6.

74. Starik, M. (1994). What is a stakeholder? Essay by Mark Starik. Pp. 89-95 of The Toronto conference: Reflections on stakeholder theory. Business & Society, 33, 82-131.

75. Starik, M. (1995). Should trees have managerial standing? Toward status for non-human nature. Journal of Business Ethics, 14, 207-217.

76. World Commission on Environment and Development. (1987). Our common future. The Brundtland report. Oxford, England: Oxford University Press.

77. Zellweger, T. M., Kellermanns, F. W., Chrisman, J. J., & Chua, J. H. (2012). Family control and family firm valuation by family CEOs: The importance of intentions for transgenerational control. Organization Science, 23, 851-868.

78. Zellweger, T. M., Nason, R. S., & Nordqvist, M. (2012). From longevity of firms to transgenerational entrepreneurship of families introducing family entrepreneurial orientation. Family Business Review, 25, 136-155.

CHAPTER 10

AN INTEGRATED SUSTAINABLE BUSINESS AND DEVELOPMENT SYSTEM: THOUGHTS AND OPINIONS

RACHEL J. C. CHEN

10.1 INTRODUCTION

Based on global environmental trends, many companies are committed to minimizing current and future negative environmental impacts while conducting their operations. To sustain a safe operational work environment, the list of priorities may include a safe performance improvement, minimizing risk and stressful situations, implementing safe and sustainable energy management systems, and incorporating know-how innovative technologies. Companies aim to utilize new technologies to establish sustainable value for their employees, customers, and stakeholders in terms of profitable growth and to minimize the carbon footprints of their operations and products. For example, creating product and process innovations aims to increase economic benefits in all aspects. Development may include packaging innovations, creative ideas to improve the sustainable quality of products, duration of economic benefits and creative ideas, and making more materials recyclable.

Companies understand the importance of monitoring and managing their environmental impacts and aim to integrate, with consistent quality control, effective reduce-reuse-recycle programs and risk preventions. By building an integrated sustainable business and development system to meet certain environmental standards, many companies are eligible to be "green" certified. Companies may consider recognizing global visions on sustainability while implementing local best practices. An integrated sustainable business and development system includes talent management, sustainable supply chain, practicing strategies of leveraging resources effectively, implementing social responsibilities, initiating innovative programs of recycling, reducing, and reusing, advancing leaders' perceptions towards sustainability, reducing innovation barriers, and engaging sustainable practices strategically. By executing sustainable consistent strategies, companies are striving to connect talented employees with committed customers to assist business growth.

10.2 AN INTEGRATED SUSTAINABLE BUSINESS AND DEVELOPMENT SYSTEM

10.2.1 TALENT MANAGEMENT

The importance of sustainable talent management has drawn the close attention of companies and corporations. By developing talented employees through engagement, retention, and leadership opportunities, associates, customers, suppliers, and shareholders will align with the corporate visions and will actively support long-term success. Aligned with company core values, many companies and businesses focus on establishing the image and brand of their employees' choice through a process of consistent assessment and in all regions culturally and effectively. Unleashing innovative ideas and globalizing company inclusion efforts will build a diverse range of leadership that will drive growth for businesses. Many companies have recognized the importance of attracting and retaining the most talented individuals and have therefore developed talent management to reduce turnover rate regionally and globally. Additionally, to enhance long-term sustainable competitiveness, it is vital to recognize opportunities among diverse supply chains and to prevent potential risks.

10.2.2 SUPPLY CHAIN

Sustainable co-values must be presented and shared with company stakeholders to achieve the key point of sustainability efforts and commercial imperatives. Investors wonder how companies evaluate the effectiveness of sustainability in the networks of supply chains. Specifically, investors and stakeholders are interested in the challenges and opportunities among corporate materiality assessments and value chains. Companies are leveraging immense resources to subscribe to the green practices that can improve business bottom lines and refine their brand images.

10.2.3 LEVERAGE RESOURCES

More small and medium companies are enticed by financial incentives to embrace applications that offer innovative solutions that will impact their daily operations and sustain positive cash flows. Those companies are also facing unforeseeable resource pressures and financial constraints while undertaking with capital improvements. Initiating proactive sustainability efforts may serve as a company's road map to distinguishing itself from its competitors and to connect with the beliefs of its customers and the expectations of its stakeholders. While facing the challenges of corporate social responsibility, embedding a sustainability culture into the company's core-values will not only reduce its operational footprints and produce significant efficiency in all aspects but will also improve economic benefits.

10.2.4 SOCIAL RESPONSIBILITY

Communities will gain jobs and economic benefits through corporate volunteerism and financial giving, and the enriched relationships between the business and public sectors will sustain thriving communities by advancing sustainable initiatives, supporting charitable activities, and civic enhancement. Various corporations provide financial donations to impact-driven communities and non-profit organizations that are committed to sustainability and that support corporate missions and values directly or indirectly, such as encouraging the engagement of residents and employees, promoting education, reusing, reducing, and recycling.

10.2.5 RECYCLE, REDUCE, AND REUSE

Recycling produces many benefits, including economic benefits, energy efficiency, primary resource efficiency, presentation of a positive image, and reducing greenhouse emissions and the carbon footprint of products. Costs for collecting and processing recycled materials must be offset to meet the bottom line or even to produce sufficient revenues. An affordable and effective collection system must be addressed in specific regions to keep pace with the trends and issues in the recycled market and to meet the needs of sorting techniques with the support of customer awareness and commitment. Corporations may consider including innovative programs that will sustain their business core values and leverage their business objectives among employees, suppliers, customers, and communities.

10.3 EXAMPLES OF INNOVATION BARRIERS

Examples of innovation barriers ranked by managerial individuals include lack of effective communication within the corporation, lack of sufficient customer input, lack of sufficient creative ideas, lack of support from top management, lack of reward for individual sustainable efforts to encourage continued engagement, lack of adequate performance measurement, and insufficient implementation of lessons learned to facilitate future action plans. Those innovative programs will focus on promoting sustainability concepts of reduce, reuse, and recycle actions, retaining high quality associates, managers, and leaders, sustaining leaders of boards of directors and members who contribute to the visions and missions of the company, establishing wellness programs, and rewarding centers that lead the sustainable business implementations to the next level of success.

10.3.1 PERCEPTIONS AND ATTITUDES OF CORPORATE LEADERS

Accenture reported the findings of surveying the perceptions and attitudes of corporate leaders toward sustainability efforts. Key barriers identified by corporate leaders regarding their hesitance in adopting and implementing sustainability include lack of government incentives (30%), additional cost (43%), lack of capacity of measuring sustainability efforts (31%), and underestimating what one company can do to prevent climate change (29%). Approximately 66% of the respondents see sustainability as a worthy investment, while 34% responded that sustainability efforts generate more cost and bring minimum to no benefits on return of investment. Approximately 49% of all respondents indicated that sustainability will generate positive benefits in reputation and trust, greater brand (41%), and lower cost (42%). Specifically, managerial associates and leaders have stated that the most common areas for sustainability initiatives are product development (44%), talent sustainability (47%), and reducing electricity usage and green initiatives (51%) [1].

Companies that were recognized for their sustainability commitment and implementation financially outperform their peers during the period of economic recession. Sustainability efforts could be woven into corporate key cores and utilize green initiatives as part of strategic approaches that can fulfill corporate social responsibility and increase company profits. Companies aim to develop strategic sustainability visions by prioritizing their core values, such as customer loyalty, reductions of energy and resource consumptions, employee retention, talent management, and evaluating greenhouse gas production and the impacts of waste generation. Governments play a vital role in motivating corporations to incorporate sustainability into their business plans and tax reduction. Through tax laws, financial supports, and public policy requirements, private sectors will benefit while making sustainability part of their core values.

According to the MIT Sloan Management Review (MIT SMR) and the Boston Consulting Group (BCG), companies that adopted and modified their business models due to the outcomes of sustainability opportunities are profiting and are recognized as "Sustainability-Driven Innovators". Approximately 2600 executives and managers of companies across the continents were surveyed and asked about their thoughts and experiences regarding sustainability efforts by their companies. The respondents indicated that more than 50% of the companies studied have changed their business models and incorporated more sustainability mindsets into various disciplines, including product development, strategic growth, marketing, supply chain, and energy efficiency [2].

Those Sustainability-Driven Innovators enjoy financial benefits due to sustainability opportunities and focus on increasing their market shares, increasing energy efficiency, and establishing greater image and brand. By working cooperatively with corporate stakeholders and involving customers while driving sustainability as main objectives of their business focus, many Sustainability-Driven Innovators have executively included sustainability within their operation and management agenda. More than half of the respondents recognized the importance of obtaining support from top-management, collaborating with customers, and implementing innovative strategies into daily company sustainability activities. Sustainability cannot only be seen as an idea, but it can result in significant financial rewards.

All companies that benefit from sustainability are aware of the importance and reap the benefits of incorporating sustainability as one of the main keys of organizational cultural, integrating sustainability activities into the corporate business models, generating top management support towards sustainability initiatives, investigating customer willingness to pay for a higher rate while involving sustainability issues and environmental conservations, and seeking sustainable internal and external supports among businesses, organizations, customers, public sectors, and individuals. Various businesses have recognized the value of incorporating social media into management, organizational communication, marketing, customer support, public relations, bridging corporate communications, and green brand addresses. Using social media wisely to deliver meaningful messages that imparts the companies' main core value to stakeholders and the target audience will launch a market that engages strategies and conveys corporate social responsibility strategies to make positive contributions at all aspects. To advance conversations and discussions among stakeholders, businesses, communities, and individuals, social media provides the platform for sustainability-focused organizations to further their engagements in all respects.

10.3.2 SUSTAINABLE IMPLEMENTATION: A HOTEL CASE

Imagine what future hotels would look like if sustainable and innovative technologies were adopted? The implementation of the reuse of property buildings includes the use of insulation on roofing, the reuse of recycled materials during a construction process, the use of LED lighting, and the reuse of water for property plant irrigation and house cleaning. In the lobby area, the use of natural lighting and ventilation and of reused lighting fixtures may reduce the cost of installing unnecessary lighting. A heat recovery system may generate cool air in the hallway. The use of a Radio Frequency Identification Technology (RFID) key card system, a CCTV fire alarm, and reused lighting fixtures can all assist the property in saving operational costs in the hallway areas. Using reclaimed solid hardwood to install built-in desks and floors, installing fixed windows for sound insulation, and implementing a linen reuse program will save operational costs. Using water-saving taps with aerators and Siphonic jet toilet flushing, installing eco-certified organic bathroom amenities, using a waste segregation program, and implementing a towel reuse program will reduce the cost of operations. In the ballroom and conference rooms, the use of eco-friendly cleaning materials and insulated double-glazed glass for windows will increase energy efficiency and reduce waste.

Hotels aim to educate their employees regarding the importance of changing their daily behavior. Hotels have also initiated programs to reduce negative environmental impacts. For example, the Hyatt Regency initiated cell phone recycling programs, reduced water volumes for sinks and toilets, installed energy efficient light bulbs, and collected plastic bags for recycling from their employees while offering a grocery bag made of recycled materials to encourage their associates to shop green. The Hyatt Earth Training program has trained more than 35,000 employees regarding the green innovation and changed their behavior to embrace a green and greater quality of life. More than 88% of Hyatt hotels recycle plastic, glass, aluminum and paper [3]. To enhance the innovative sustainability movement to the next level of success, many companies in the hospitality and tourism industry have been focusing on engaging their associates and customers in environmental initiatives and empowering numerous team members to make positive changes and to create notable positive environmental impacts. A list of examples includes:

- Companies and many public places provide filtered water stations to reduce the waste involved in using plastic water bottles.
- Companies have proved the advantages of using 100% recycled paper to support the awareness of earth day.
- During a conference, placing note pads and pens in a central region for conference attendees.
- Making the right decisions to eliminate the use of disposable products and to set the room temperature is an efficient approach to reducing unnecessary waste.
- Many hotels are implementing innovative ideas, including using AC condensate water to sustain their roof-top veggie/herb garden and providing plug-in power for hybrid vehicle guests.
- Many properties provide recycling bins at various spaces as a green service.
- Printing on-site and on recycled paper is friendlier to the environment.
- Properties offer communication channels that encourage guests, employees, and stakeholders to explore their creative ideas around sustainability to enhance a greener image of hotel operations.
- Sharing ideas and thoughts through social media to demonstrate how properties initiate their innovative ideas through sustainable

efforts integrates with their customers to add opportunities for implementing Corporate Responsibility (CR).

- The creation of a menu could be impacted by guests' eating habits. Understanding customers' behaviors can benefit the bottom line of the properties and restaurants.
- Through innovative implementation in the hospitality and tourism industries, many major companies have experienced and expected positive increases in their profit margins and success in waste management, energy effectiveness, water conservation, and savings in operations.
- Utilizing locally sourced and grown products and featuring local flavors will reduce transportation emission and transportation costs.
- While heading to a meeting, shipping less during the trip will reduce fuel, energy, and paper waste.

Overall, by incorporating achievable sustainable initiatives through integrating every facet across all operational divisions among companies, the positive impacts of environmental contributions are measurable and sustainable in the communities. Such companies as Hyatt, Marriott, and Hilton demonstrate their missions of sustainability efforts by creating a corporate culture that aims to pursue social responsibility and environmental awareness that empowers their employees at lodging properties, for example, to initiate and identify opportunities to increase energy efficiency, reduce water consumption and waste, and that encourages them to make positive green changes in their own dwelling places.

Many properties feature the importance of constructing sustainable buildings that are able to incorporate innovative technology and greener operations by including environmentally friendly materials. Properties are committed to utilizing and purchasing more environmentally preferable materials, including recycled plastic bottles of shampoo and lotion, recycled plastic, reusable cloth laundry bags, recycled carpeting, zero-VOC paint, and locally sourced menu items. Properties also have installed thousands of LED light bulbs in the past three years. Restaurants offer antibiotic/hormone-free beef hamburgers and cage free eggs. For instance, using EcoLab cleaning detergents and products has reduced the waste of millions of gallons of water and of natural gas by hundreds of thousands of themes. Guests have been encouraged to reduce the frequency of linen changes and are informed how their behaviors have helped to preserve the natural environment by reducing the use of chemicals, water, and energy. Assessing energy consumption patterns among properties, understanding the renewable energy potentials, and undergoing energy audits can assist companies to prepare various energy efficiency programs across public and private sectors.

3.3. SUSTAINABLE FOOD AND AGRICULTURE: A CASE OF SUSTAINABLE FOOD

Organic agriculture and food production may minimize adverse environmental impacts, increase the level of biodiversity, decrease soil and plant pollution caused by pesticides, lower carbon emission, and sustain energy efficiency. The United States Department of Agriculture (USDA) has approved a set of standards that assure consumers that foods labeled "organic products" are appropriately processed and certified [4]. Sustainability implementations on livestock and plant production and packaging can generate various social, environmental, and economic benefits. Most recently, the trends of using locally produced products, such as those produced in the backyard garden, may reduce greenhouse gas emission, air pollution, gasoline cost, noise pollution, congestion, and accidents while compared to the cost of using food products that have travelled long distances to get to the shops, such as imported food related products. Additionally, local farmers may benefit from the increased demand from their local consumers.

According the U.S. National Restaurant Association [5], the top 10 most important trends identified by interviewed chefs included locally sourced meats and seafood, locally grown products, healthful meals for children, hyper-local sourcing, sustainability, children's nutrition, gluten-free/allergy-conscious food, locally-produced wine and beer, sustainable seafood, and wholegrain items in children's meals. More individuals and organizations agree with the terms of "sustainable food" as safe, healthy and nutritious. Additionally, they believe in respecting biophysical ecosystems and in recognizing environmental limits. They also aim to assist in supporting rural economies, farmers, and sustainable local products. Through certain lifestyles changes, these committed groups have positively affected the vegetarian movement, protecting the natural environment and focusing on animal welfare.

4. SUMMARY

Companies focus on improving their operational processes to support environmental systems, reduce water and energy consumption, and divert reusable waste to increase economic benefits. Many corporations and companies have established programs to reduce negative environmental impacts, reduce operational costs, and implement compliance strategies to perform consistently in significant ways. Environmental sustainability requires a long-term commitment and endeavors that consist of innovative implementation and conscious awareness with a willingness to change behaviors. Companies foresee the need to set achievable goals for their annual green initiatives to reduce gas emissions, waste, water consumption, and energy use on a daily operational basis. By tracking the effectiveness of progress, those committed companies continue to identify opportunities for continuously innovative development and improvement. Overall, a comprehensive list of sustainability efforts may include, but are not limited to, the following:

- *Secure a sustainable business* may include (1) starting a sustainable business plan that aims at business innovation and growth; (2) communicating with associates and stakeholders to explain why and how green initiatives can make a recognized brand and workplace; (3) challenging each employee and executive to re-think their roles within business sustainability; and (4) connecting the ties between bottom line and sustainability strategies.
- *Energy Efficiency.* Presenting and pre-assessing the benefits of focusing on energy efficiency across the corporation and how energy can boost the business bottom line.
- *Engage employees* such as (1) compiling a plan that encourages employees to be engaged and committed to corporate social

responsibility; and (2) rewarding corporate divisions and individuals who walk extra miles to promote and implement green deeds.

- *Community involvement and partnership.* Engaging and providing incentives to encourage communities to be part of the green initiatives through financial award systems.
- *Support and ideas from stakeholders.* Welcoming stakeholders' innovative ideas and utilizing social media to provide more timely communications among all parties to increase effective engagements.
- *Sustainable Leadership.* Sustaining talented associates and leaders of the corporation will likely reduce the cost of human resources. Companies need to sustain a healthy work culture that empowers their employees to resolve issues on a daily basis.
- *Stakeholder Engagement.* Examples consists of (1) using regular communications at horizontal and vertical levels to ensure that the overall visions and sustainable business strategies are well presented to internal and external partners; (2) challenging companies to tell their stories about their sustainability ideas and implementation; and (3) incorporating daily sustainability efforts into associates' performance metrics to meet corporate social reasonability objectives [6,7].

Collaboration between public and private sectors can launch effective sustainable projects that focus on reducing carbon footprints, reducing the use of fossil fuel, increase the use of renewable energy, increase corporate competitiveness, and improve corporate images. Utilizing environment-friendly technologies will lead societies to a smarter place that can foster greater life styles and build sustainable societies for future generations. The issues and trends of climate change, deforestation, biodiversity loss, and water shortages have attracted attention among nations. More corporate leaders have established goals that are related to ecosystem services to embrace aspirational visions that foster zero emissions, water and carbon neutrality, and minimization of adverse impacts on biodiversity.

Identifying and implementing innovative approaches to resolve sustainability challenges of the present and future can assist in minimizing carbon footprints and maximizing economic benefits for companies. To ensure the validity and effectiveness of implementing sustainable business models, developing measurable matrices that can assess and evaluate the pre and post analyses of sustainability efforts will assist corporations to sustain their bottom lines and to maximize their return on investment. More executive individuals have captured the interest of promoting innovation across companies that are actively participating in the path of an aligned corporate culture towards their business sustainability.

Identifying achievable objects by converting innovative ideas into actual implementable projects can assist dialogs that can energize supply chain stakeholders, incorporate customer needs, and engage employees in the direction of aligning corporate core values. Identifying business sustainability benefits and pre-measuring business derailleurs can favorably position companies in the journey of global economic recovery. Social media engagement can be used to strengthen a company's network and to complement the company's sustainable business strategy by broadcasting messages to prospective markets and customers. Many customers have indicated that they are interested in participating in co-developing products and making co-decisions aligned with their interested companies. By sustaining a high bar, extending corporate visions toward sustainable business, motivating influential partnerships, equipping associates strategically, and capturing the impacts of sustainability on society, achievable mitigations of sustainability efforts may be structured constructively in a timely manner.

As mentioned previously, companies are leveraging immense resources to subscribe to green practices that can improve business bottom lines and refine their brand images. While facing the challenges of corporate social responsibility, embedding a sustainability culture into the company's core values will not only reduce the company's operational footprints and produce significant efficiencies in all aspects, but will also improve economic benefits. Communities will gain jobs and economic benefits through corporate volunteerism and financial giving. Also, the enriched relationships between the business and public sectors will sustain thriving communities by advancing sustainable initiatives, supporting charitable activities, and promoting civic enhancement. This Special Issue aims to discuss strategy frameworks from a sustainable business and development perspective. Scholars across the continents contribute to this issue by submitting comprehensive reviews, case studies or research articles, presenting a variety of methodological approaches, and offering suggestions for further implementations.

REFERENCES

1. Blonkowski, N.; Jones, D.; Naik, S.; Raman, S. The Value of the Sustainable Supply Chain: What Do Consumers Think? Available online: http://www.accenture.com/SiteCollectionDocuments/PDF/Accenture-The-Value-of-the-Sustainable-Supply-Chain.pdf (accessed on 6 July 2014).
2. Kiron, D.; Kruschwitz, N.; Rubel, H.; Reeves, M.; Fuisz-Kehrbach, S.K. Sustainability's Next Frontier: Walking the Talk on the Sustainability Issues That Matter Most. Available online: http://sloanreview.mit.edu/projects/sustainabilitys-next-frontier/ (accessed on 2 February 2014).
3. HYATT. Meet and Be Green Commitment. Available online: http://www.hyattmeetings.com/Green-Details.asp (accessed on 30 January 2014).
4. USDA Organic. Labeling Organic Products. Available online: http://www.ams.usda.gov/AMSv1.0/getfile?dDocName=STELDEV3004446 (accessed on 28 March 2014).
5. National Restaurant Association. What's Hot in 2012 chef survey shows local sourcing, kids' nutrition as top menu trends? Available online: http://www.restaurant.org/News-Research/News/What-s-Hot-in-2012-chef-survey-shows-local-sourcin (accessed on 1 May 2014).
6. Sloan, P.; Legrand, W.; Chen, J.S. Sustainability in the Hospitality Industry: Principles of Sustainable Operations, 2nd ed. ed.; Routledge: London, UK, 2013.
7. Parsa, H.G.; Segarra-Ona, M.; Jang, S.C.; Chen, R.J.C.; Singh, A.J. Special Issue on Sustainable and Eco-Innovative Practices in Hospitality and Tourism. Cornell Hospit. Q. 2014, 55. Article 5.

APPENDIX

USDA GUIDELINES FOR LABELING WINE WITH ORGANIC REFERENCES

KEY

PDP - (PRINCIPAL DISPLAY PANEL) That part of the label that is most likely to be displayed, presented, shown, or examined under customary conditions of display for sale.

IP - (INFORMATION PANEL) That part of the label of a packaged product that is immediately contiguous to and to the right of the principal display panel as observed by an individual facing the principal display panel, unless an other section of the label is designated as the information panel because of pack age size or other at tributes (e.g., irregular shape with one usable surface).

OP - (OTHER PANEL) Any panel other than the principal display panel, information panel, or ingredient state ment.

IS - (INGREDIENT STATEMENT) The list of ingredients contained in a product shown in their common and usual names in the descending order of predominance.

Grapeful

100%
Organic
Wine

California
Cabernet Sauvignon.
100% Organically Grown Grapes

13.2% Alc./Vol.
750 ml

Grapeful

100%
Organic
Wine

INGREDIENTS: ORGANIC GRAPES

PRODUCED & BOTTLED BY:
Richard John Vineyards
Napa Valley, CA

CERTIFIED ORGANIC BY:
Evanchec Organic Certifiers

GOVERNMENT WARNING: (1) ACCORDING TO THE SUR-
GEON GENERAL, WOMEN SHOULD NOT DRINK ALCOHOLIC
BEVERAGES DURING PREGNANCY BECAUSE OF THE RISK OF
BIRTH DEFECTS. (2) CONSUMPTION OF ALCOHOLIC BEV-
ERAGES IMPAIRS YOUR ABILITY TO DRIVE A CAR OR OPERATE
MACHINERY, AND MAY CAUSE HEALTH PROBLEMS.

GUIDELINES FOR LABELING WINE AS "100% ORGANIC"

This document contains a sample label. It should be used as guidance relating to the National Organic Program (NOP) regulations at 7 CFR part 205. To view these regulations in their entirety, please visit the United States Department of Agriculture's website at www.ams.usda.gov/nop. This sample com plies with the Federal Alcohol Administration Act, the Alcohol Beverage Labeling Act and the NOP.

When labeling your product as "100% Organic," it must contain 100 percent organically produced ingredients, not counting added water and salt. You should also consider the following points in designing your label.

BRAND NAME/CLASS & TYPE (OPTIONAL)

The phrase "100% Organic" may be used to modify the brand name and/ or class and type statement on the PDP, the IP, or the OP.

SULFITE STATEMENT

"100% Organic" products cannot use added sulfites in production. Therefore, since no added sulfites are present in the finished product, the label may not require a sulfite statement. In these cases, a lab analysis is necessary to verify that the wine contains less than 10 ppm of sulfites.

INGREDIENT STATEMENT (REQUIRED)

Products certified as "100% Organic" must show a complete ingredient statement. The term "organic" may be used to identify the specific ingredients. Water and salt included as ingredients may not be identified as "organic."

Organic
Grapeful
California
Cabernet Sauvignon

97% Organic

13.2% Alc./Vol.
750 ml

Organic
Grapeful
INGREDIENTS: ORGANIC CABERNET
SAUVIGNON GRAPES, TANNAT GRAPES
PRODUCED & BOTTLED BY:
Richard John Vineyards
Napa Valley, CA
CERTIFIED ORGANIC BY:
Evanchec Organic Certifiers

GOVERNMENT WARNING: (1) ACCORDING TO THE SUR-
GEON GENERAL, WOMEN SHOULD NOT DRINK ALCOHOLIC
BEVERAGES DURING PREGNANCY BECAUSE OF THE RISK OF
BIRTH DEFECTS. (2) CONSUMPTION OF ALCOHOLIC BEV-
ERAGES IMPAIRS YOUR ABILITY TO DRIVE A CAR OR OPERATE
MACHINERY, AND MAY CAUSE HEALTH PROBLEMS.

CERTIFICATION STATEMENT (REQUIRED)

"Certified Organic by ---" or a similar phrase must be listed below the name and address of the producer, bottler, importer, etc. This statement must be on the IP and may include the agent's business address telephone number, or internet address.

USDA ORGANIC SEAL (OPTIONAL)

The USDA Organic Seal may be placed on the label of a product that is certified as "100% Organic." This seal may only appear on the PDP or OP.

CERTIFYING AGENT SEAL (OPTIONAL)

The seal of a USDA-accredited certifying agent may be placed on the label of a product that is certified as "100% Organic." This seal may only appear on the PDP or OP.

GUIDELINES FOR LABELING WINE AS "ORGANIC"

This document contains a sample label. It should be used as guidance relating to the National Organic Program (NOP) regulations at 7 CFR part 205. To view these regulations in their entirety, please visit the United States Department of Agriculture's website at www.ams.usda.gov/nop. This sample complies with the Federal Alcohol Administration Act, the Alcohol Beverage Labeling Act and the NOP.

When labeling your product as "Organic," it must contain at least 95 percent organically produced ingredients, not count ing added water and salt. In addition, your product must not contain added sulfites and may contain up to 5 percent nonorganically produced agricultural ingredients which are not commercially available in organic form and/or other substances, including yeast, allowed by 7 CFR 205.605. You should also consider the following points in designing your label.

BRAND NAME/CLASS & TYPE (OPTIONAL)

The term "organic" may be used to modify the brand name and/or class and type statement on the PDP, the IP, or the OP.

SULFITE STATEMENT

"Organic" products cannot use added sulfites in production. Therefore, since no added sulfites are present in the finished product, the label may not require a sulfite statement. In these cases, a lab analysis is necessary to verify that the wine contains less than 10 ppm of sulfites.

PERCENTAGE STATEMENT (OPTIONAL)

The phrase "X% Organic" or "X% Organic Ingredients" may be included on the labeling. Such statements may be included on the labeling. Such statements may appear on the PDP, the IP, or the OP.

INGREDIENT STATEMENT (REQUIRED)

Products certified as "organic" must show a complete ingredient statement. The term "organic" may be used to identify the specific ingredients. Water and salt included as ingredients may not be identified as "organic."

CERTIFICATION STATEMENT (REQUIRED)

"Certified Organic by ---" or a similar phrase must be listed below the name and address of the producer, bottler, importer, etc. This statement must be on the IP and may include the agent's business address, telephone number, or internet address.

CERTIFYING AGENT SEAL (OPTIONAL)

The seal of a USDA-accredited certifying agent may be placed on the label of a product that is certified as "organic." This seal may only appear on the PDP or OP.

USDA ORGANIC SEAL (OPTIONAL)

The USDA Organic Seal may be placed on the la bel of be placed on the label of a product that is certified as "organic." This seal may only appear on the PDP or OP.

GUIDELINES FOR LABELING WINE AS "MADE WITH ORGANIC INGREDIENTS"

This document contains a sample label. It should be used as guidance relating to the National Organic Program (NOP) regulations at 7 CFR part 205. To view these regulations in their entirety, please visit the United States Department of Agriculture's website at www.ams.usda.gov/nop. This sample complies with the Federal Alcohol Administration Act, the Alcohol Beverage Labeling Act and the NOP.

When labeling your product as "Made with Organic Ingredients" (or a similar phrase), it must contain at least 70 percent organically produced ingredients, not counting added water and salt. In addition, wine may con tain added sulfites (in accordance with 7 CFR 205.605) and may contain up to 30 percent nonorganically produced agricultural ingredients and/or other substances, including yeast, allowed by 7 CFR 205.605. You should also consider the following points in designing your label.

Grapeful

California
Cabernet Sauvignon

Made with Organic Grapes

13.2% Alc./Vol.
750 ml

Grapeful
75% Organic Ingredients

INGREDIENTS:
ORGANIC CABERNET
SAUVIGNON GRAPES,
PRODUCED & BOTTLED BY: TANNAT GRAPES.
Richard John Vineyards   **CONTAINS**
Napa Valley, CA   **SULFITES**

CERTIFIED ORGANIC BY:
Evanchec Organic Certifiers

EVANCHEC

GOVERNMENT WARNING: (1) ACCORDING TO THE SUR-
GEON GENERAL, WOMEN SHOULD NOT DRINK ALCOHOLIC
BEVERAGES DURING PREGNANCY BECAUSE OF THE RISK OF
BIRTH DEFECTS. (2) CONSUMPTION OF ALCOHOLIC BEV-
ERAGES IMPAIRS YOUR ABILITY TO DRIVE A CAR OR OPERATE
MACHINERY, AND MAY CAUSE HEALTH PROBLEMS.

"MADE WITH ORGANIC ---" STATEMENT (OPTIONAL)

The phrase "Made with Organic ---" (specified ingredients or food groups) may be included on the labeling. Such statements may appear on the PDP, the IP, or the OP.

INGREDIENT STATEMENT (REQUIRED)

Products certified as "Made with Organic Ingredients" must show a complete ingredient statement. The term "organic" may be used to identify the specific ingredients. Water and salt included as ingredients may not be identified as "organic."

"X% ORGANIC" STATEMENT (OPTIONAL)

The phrase "X% Organic" or "X% Organic Ingredients" may be included on the labeling. This statement may appear on the PDP, the IP, or the OP.

CERTIFICATION STATEMENT (REQUIRED)

"Certified Organic by ---" or a similar phrase must be listed below the name and address of the producer, bottler, importer, etc. This statement must be on the IP and may include the agent's business address, telephone number, or internet address.

CERTIFYING AGENT SEAL (OPTIONAL)

The seal of a USDA-accredited certifying agent maybe placed on the label of a product that is certified as "Made with Organic Ingredients." This seal may only appear on the PDP or OP.

Grapeful

California
Cabernet Sauvignon

13.2% Alc./Vol.
750 ml

Grapeful
50% Organic Ingredients

PRODUCED & BOTTLED BY:
Richard John Vineyards
Napa Valley, CA

CERTIFIED ORGANIC BY:
Evanchec Organic Certifiers

INGREDIENTS:
ORGANIC CABERNET
SAUVIGNON GRAPES,
TANNAT GRAPES,

**CONTAINS
SULFITES**

GOVERNMENT WARNING: (1) ACCORDING TO THE SUR-
GEON GENERAL, WOMEN SHOULD NOT DRINK ALCOHOLIC
BEVERAGES DURING PREGNANCY BECAUSE OF THE RISK OF
BIRTH DEFECTS. (2) CONSUMPTION OF ALCOHOLIC BEV-
ERAGES IMPAIRS YOUR ABILITY TO DRIVE A CAR OR OPERATE
MACHINERY, AND MAY CAUSE HEALTH PROBLEMS.

THE LABEL MAY NOT SHOW:
* THE USDA ORGANIC SEAL

GUIDELINES FOR CLAIMING THAT WINE
CONTAINS SOME ORGANIC INGREDIENTS

This document contains a sample label. It should be used as guidance relating to the National Organic Program (NOP) regulations at 7 CFR part 205. To view these regulations in their entirety, please visit the United States Department of Agriculture's website at www.ams.usda.gov/nop. This sample complies with the Federal Alcohol Administration Act, the Alcohol Beverage Labeling Act and the NOP.

When claiming that your product contains some organic ingredients, it may contain less than 70 percent organically produced ingredients, not counting added water and salt. In addition, the product may contain over 30 percent nonorganically produced agricultural ingredients and/or other substances, including yeast, without being limited to those in 7 CFR 205.605. Your label should also consider the following points in designing your label.

INGREDIENT STATEMENT (REQUIRED)

Products that wish to identify some ingredients as "organic" must show a complete ingredient statement. The term "organic" may be used to identify the specific ingredients. Water and salt included as ingredients may not be identified as "organic."

"X% ORGANIC" STATEMENT (REQUIRED)

The phrase "X% Organic" or "X% Organic Ingredients" must be included on the labeling when organically produced ingredients are identified in the ingredient statement. This statement must appear on the IP.

AUTHOR NOTES

CHAPTER 1

Acknowledgments
We would like to acknowledge the collaboration of Ana Elisa Rodríguez Carril, Gamma representative, and Anxo Estévez Carril, whose effort made this study possible.

Research Grants
This study was made possible thanks to the financial support of the research project "Ecological Footprint for IMAPS (Integrated Management of risks and Environment in Port Cities), leaded by the Port of Gijón.

CHAPTER 2

Acknowledgments
This work has been funded by EPSRC within the CCaLC and CSEF projects (grants no. EP/F003501/1 &EP/K011820/1). This funding is gratefully acknowledged.

CHAPTER 3

Conflict of Interest Statement
The authors declare that the research was conducted in the absence of any commercial or financial relationships that could be construed as a potential conflict of interest.

Acknowledgments
This work is part of the project "Oenological microbiota: selection to identify the wine character and to improve the competitiveness of Montepulciano d'Abruzzo wineries" supported by grant from Cassa di Risparmio di Teramo.

CHAPTER 4

Acknowledgment
Fundação para a Ciência e a Tecnologia (FCT) – PhD grant SFRH/ BD/ 31653/ 2006, by financial support and Quinta da Casaboa, Catapereiro and Herdade da Mingorra where the studies were conducted.

CHAPTER 5

Acknowledgments
This study has been supported by the European Aluminium Foil Association (E.A.F.A.). The authors would like to thank Christian Bauer of E.A.F.A., Cyril Barioz of Amcor and Gerald Rebitzer of Amcor, for their valuable inputs.

CHAPTER 8

Acknowledgment
I would like to thank Mario Annunziata for the design and manufacture of the labels.

CHAPTER 9

Acknowledgment
This research was conducted with the following undergraduate students at UCLA: Yousuf Anvery, Sachin Goel, Shilpa Hareesh, John Hogan, Antonio Menchaca, Roxana Ramirez. We thank them for their essential input.

Declaration of Conflicting Interests
The author(s) declared no potential conflicts of interest with respect to the research, authorship, and/or publication of this article.

Funding
The author(s) disclosed receipt of the following financial support for the research, authorship, and/or publication of this article: We would like to thank the Conseil R.gional de Champagne-Ardenne and the UCLA Center for European and Eurasian Studies for their support.

INDEX